An idea for a theatre ecology

Manchester University Press

An idea for a theatre ecology

Methods, theories, histories and practices

Carl Lavery

MANCHESTER UNIVERSITY PRESS

Copyright © Carl Lavery 2025

The right of Carl Lavery to be identified as the author of this work has been asserted in accordance with the Copyright, Designs and Patents Act 1988.

Published by Manchester University Press
Oxford Road, Manchester, M13 9PL
www.manchesteruniversitypress.co.uk

British Library Cataloguing-in-Publication Data
A catalogue record for this book is available from the British Library

ISBN 978 1 5261 8892 2 hardback

First published 2025

The publisher has no responsibility for the persistence or accuracy of URLs for any external or third-party internet websites referred to in this book, and does not guarantee that any content on such websites is, or will remain, accurate or appropriate.

EU authorised representative for GPSR:
Easy Access System Europe – Mustamäe tee 50, 10621 Tallinn, Estonia
gpsr.requests@easproject.com

Typeset
by Cheshire Typesetting Ltd, Cuddington, Cheshire

To Melanie – for everything, really

To Immanuel, Inez and Saul – just to prove that I do actually go to work and manage to escape the lazy pull of the Western Baths from time to time

To Mali – for the love and friendship

Contents

Polemos	*page* ix
Acknowledgements	xvii
Introduction: an idea for a theatre ecology	1
1 Contexts (questioning methods)	25
2 Theory (theatricalising ecology)	78
3 Model (a lexicon for theatre ecology)	106
4 Concept (ecologising theatre)	122
5 History (Artaud's cruel ecology)	141
6 Analysis (becoming archipelagic)	182
Afterlude: double cut/frozen wave	207
Glossary	213
Bibliography	219
Index	242

Polemos

Constrained by the demands of a marketplace that privileges products and outputs and belonging to a globalised university culture that 'scans horizons' for economic opportunities, any academic text that harbours progressive ecological ambitions, as this one does, calls out for some kind of justification. How to produce something that would not contribute, by the simple fact of its existence, to an anti-ecological culture of research excellence, in which cost, not value, is the *sine qua non*?[1]

Might silence have been better?

One can never be sure.

And yet ...

Wagering on the optimistic possibility that this text has the capacity to escape a metrics of capture, it is imperative – for the sake of polemic, at the very least – to clarify why I have written this book, the first volume in a triptych of sorts about theatre's relationship with the earth. In keeping with Félix Guattari's notion of 'transversality',[2] the line that simultaneously negates and creates differences, I regard this book as fulfilling an activist agenda of sorts, even though any detailed description of eco-activist performance practices is noticeable by its absence.[3] Whereas eco-activist artists, to speak broadly, seek to change behaviour patterns through strategic acts of aesthetico-political intervention, my approach is both more immanent and oblique. The aim is to demonstrate how a 'theatricalised theatre' can allow for alternative modes of relating to the earth through the creation of autonomous, aesthetic affect. In the words of Gilles Deleuze and Félix Guattari's *What Is Philosophy?*, I am interested in speculating on how theatre, through the 'cruel' spell that it inflicts on bodies and minds, can contribute to the construction of both a 'new earth' and a 'people to come' (Deleuze and Guattari, 1994: 218).

Such enthusiasm for the affective power of the theatrical event would appear to have little in common with accepted models of eco-activism in theatre and performance. Like all activism, eco-activism does not have the luxury of speculating on the aesthetic disorientations and dizzying becomings that concern me in this book. Its terrain is one of pragmatic choices, even when, as Gerald Raunig argues in the sophisticated *Art and Revolution: Transversal Activism in the Long Twentieth Century* (Raunig, 2007), activist artists do not simply aim to politicise the aesthetic but instead respect its 'singularity', its unique ways of operating (Raunig, 2007: 264–5).

Yet for all his concern with singularity, the performances that Raunig focuses on are nevertheless easily 'read' as political, inserting themselves within recognisable contexts and looking to bring about concrete outcomes (Raunig, 2007: 18).[4] This is hardly surprising. In their admirable urge to change the world, eco-activists are inevitably grounded in the actual, in manufacturing efficacious interventions in the here and now. Paradoxically, though, it is precisely here, in the very strength of its commitments, that activism reveals its shortcomings. For by adhering so closely to a logic of cause and effect, in setting itself clear targets and goals, there is always a danger that activism may finish by replaying the same instrumentalist ways of seeing and being that have done so much to damage the world in the first place.[5] As Franco Berardi proposes, activism's *indubitable* allegiance to Leninist notions of voluntarism veils the darker motivations and anxieties that haunt its mobilisation of creative energies – a danger that provokes Berardi to ask if we 'should not free ourselves from the thirst for activism that led the twentieth century to the point of catastrophe and war' (Berardi, 2011: 37).[6]

The caution inherent in Berardi's question received its (negative) confirmation in an abstract I was sent for a plenary talk at the Green Arts Conference in Edinburgh in 2019, entitled '"Art for Art's Sake Is the Philosophy of the Well-Fed": Creativity in Our Times'. The title of the talk proclaimed that, at a time of environmental crisis and emergency, satiated art ought to be rejected in favour of hungrier, leaner works whose supposed target was to 'bring about a fairer, more sustainable world'. Leaving aside the thorny questions of whether there can ever be such a thing as 'art for art's sake', there are important theoretical and pragmatic issues obscured by the imperative asserted here. Not only does such a stance fail to interrogate the performative logic inherent in the idea of declaring an 'emergency', or to reflect on the equally problematic significance of sustainability, it also does not consider the historical failure of most militant art.[7] In fact, a backward glance at the history of agit-prop and didactic art in the twentieth century would give the lie to the type of explicitly committed work being advanced here. As any student of the historical avant-garde knows, politicised art is little able to 'resist' historical violence by seeking to compete with politics, economics and policy on their own terms. Rather art is better able to exist as what Gregory Bateson terms the '*difference that makes a difference*' when it insists on constructing experiences that are open-ended, contingent and provisional (Bateson, 1972: 459; original italics).

At this juncture, a crucial but naive question imposes itself: how can one ever expect to believe in the world, if there is no belief in art? Art is ecological to its core, as Theodor Adorno, following Immanuel Kant, insisted on in his 1970 publication *Aesthetic Theory*.[8] It exists to affirm both 'Nature' and cosmos, to provoke interest and value, to give us a 'taste' for earth.[9] What I call the immanent ecology of the artwork – the fact that it *is* terrestrial – is underscored by Elizabeth Grosz:

> Through the plane of composition it casts, art is the way that the universe most directly intensifies life, enervates organs, mobilizes forces. It is the passage from the house to the universe, from territory to deterritorialization, 'from the finite to the infinite', from the body of the living being to the universe itself. (Grosz, 2008: 24)

By suggesting that art's terrestrial potential does not reside in what it *represents*, but in how it organises and structures earthly matter, Grosz is able to advance a very different idea of mobilisation from the one proposed by activists. In Grosz's view, mobilisation is not an addendum to the artwork, an extra-artistic message that the artwork articulates. Rather, it comes from the artwork's capacity to compose images and textures that intensify bodies made out of the same terrestrial stuff. Something sensate and unspoken occurs in that 'passage' from 'house' to 'universe', a force more than an idea – the genesis of a terrestrial and cosmic attachment. Mobilisation happens when materials are allowed to exist autonomously – when, that is, they touch bodies by being themselves. In the face of such radical immanence how, then, to proceed? What kind of thinking is required to draw out the autonomous ecology of the artwork, and, as a consequence, to establish new, more generative relations with activist discourses and practices? A useful starting point presents itself in John Cage's ironic declaration, made in 1965, that the desire to 'improve the world', all too often, 'only serves make it worse'.[10]

In order for activist art to lend an ear to Cage's warning, it needs a theory and practice of criticism that would 'actively' problematise some of its most cherished attachments and assumptions. That problematisation does not mean that activism and activist art are to be rejected in any absolutist manner. Far from it! Art and theory *need activism to succeed*. The institutions that uphold our present 'military-industrial-media complex' in what some call the Anthropocene and others (with more political and historical discernment) the Capitalocene or Plantationocene will not change by themselves, and neither will they wither away.[11] In that respect, the pressure that activist organisations and artists, such as Extinction Rebellion,

Dark Mountain, Platform and Liberate Tate exert on public bodies, international companies and governments is crucial, a key component in contesting today's neoliberal hegemony. What I am proposing, then, is that if we are to resist 'the barbarism' that seems to be increasing daily, a more nuanced and generous exchange needs to take place between activism and art (Stengers, 2015).[12] This would be one that does not simplistically condemn autonomous art as a privileged affair, the work of dilettantes. But rather concurs with Guattari's position in *Chaosmosis: An Ethico-Aesthetic Paradigm* (Guattari, 1995), when he proposes that the artwork gives shelter to activism of a different order:[13]

> It is in underground art that we find some of the most important cells of resistance against the steamroller of capitalist subjectivity [...]. An ecology of the virtual is thus just as pressing as ecologies of the visual world. And in this regard, poetry, music, the plastic arts – particularly in their performance or performative modalities – have an important part to play, with their specific contribution and as a paradigm of reference in new social and analytical practices. (Guattari, 1995: 90–1)

A conceptual figure for the 'ecology of the virtual' that Guattari so fiercely insists upon is found in Jacques Derrida's notion of the 'friendly enemy' (Derrida, 1997: 33).[14] There is in this queer, deconstructive relationship a refusal of complementarity as well as 'contemporaneity', an attempt to do away with the need for an enemy to sacrifice, on the one hand, and, on the other, the equally dangerous obligation to stand 'shoulder to shoulder' with the friend (Agamben, 2009: 39–54). To declare oneself a 'friendly enemy' means always being out-of-step with the friend, never quite 'on time', never definitive nor fixed. Through this syncopated, disjunctive relation a new kind of intimacy between activism and autonomous art is intuited, one that gives birth to an alternative form of aesthetics *and* politics characterised by inclusivity *and* alterity.

Via asymmetry and anachrony, the friend and enemy share the same goal, precisely because they follow different paths. In doing so, they show the impossibility of ever arriving at some absolute *telos* or final destination, a disclosure which is ethically, existentially and ecologically vital, since the violence of western, capitalist modernity is largely predicated on being 'on time'.[15] Like the slave-owner Pozzo in Beckett's 1953 play *Waiting for Godot*, capitalist subjectivity – the one that has now been internalised on a global scale – is desperate to keep its appointments, no matter how catastrophic the time keeping, no matter how devastated and barren the earth it leaves behind. What *counts*, in *Godot*, as in capitalist modernity in general, is wholeness, self-coincidence, symmetry, ecologically exhausted ways of being that Antonin Artaud excoriated in *Pour en finir avec le jugement de*

Dieu (*To Have Done with the Judgement of God*), his 1947 radio piece written on the cusp of the great acceleration in carbon burning provoked by the petro-dollar imperialism of the Marshall Plan.

To exist as a friendly enemy is to refuse God's judgement, to accept dis-*appointment* as an ontological given, to inhabit a mobile and intermittent ground that refuses to settle or fix itself. Stacey Alaimo refers to this interstice as a 'contact zone' (Alaimo, 2008: 238), and the quantum feminist thinker Karen Barad describes it as being 'cut together-apart' (Barad, 2014: 168). In fidelity to the spatio-temporal anachrony that runs through Alaimo's and Barad's thinking, the dialogue between activism and theatricality that I have staged in this book is never confronted head on. Rather, it exists as a ghost that haunts the text, an oblique 'presence', a 'gesture' that hails and reaches out, as Rebecca Schneider and Paul Rae have it in their work on theatrical temporalities (Schneider and Rae, 2018: 13–24).

To turn to Derrida again, the interrogation implicit in that theatrical hailing centres on the simple but profound ecopolitical question par excellence: namely, '*what does it mean to live well, finally*' (Derrida, 1994: xvvii–xviii; original italics)? Today 'living well finally' does not – it cannot – only refer to a community of humans, as indeed Derrida recognises in his own late texts on animals, sovereignty and 'eating well' (Derrida, 1991: 155). It has to be supplemented with a willingness to meet and embrace the non-human that dwells within and beyond us, an *oikos*, that, to cite Jean-François Lyotard, knows nothing of laws and numbers and whose joys are inseparable from a sense of 'distress' (Lyotard, 1993a: 106). The impossible ambivalence that I am referring to when I mention the Lyotardian *oikos* is part of the disturbing doubleness of the earth, too, a non-human mode of expression that can never be stilled, made to serve a predetermined purpose or humanist end. Like theatre, the earth is an event, a reckless extravagance that resits capture by the sad economics of surplus value – a scandal to thought. For me, the ecological potential of performance is found in its capacity to produce terrestrial forms of thinking and feeling. Such an ecology is full of pathos, predicated on evocation rather than exemplification, in encouraging spectators to open themselves, in their bodies and minds, to the im/material becomings of the earth: its unseizable force, its power to experiment with itself.

To get to that earth, to reach that theatre, it seemed important to start not with a foreword as such, but with something *gestic*, a paradoxical *polemos* whose violence was directed at a lesser violence.[16] The hope in such a *polemos* is double: both to instigate a new dialogue between different modes of activism *and* to register and contest the sense of derangement that so many of us feel in the 'ruins' of the 'globally excellent' university – its toxic ecology.[17]

Most of this text was written during the Covid-19 pandemic, and its publication was substantially delayed because of it. Reading it now, there are certain things I could add, but I have decided to keep the text as it is. It provides a document of the times. And I still believe, strongly, in the central idea advanced: namely, that there is an urgent need to reposition the lenses through which the relationship between theatre and ecology has been habitually viewed.

Notes

1 In making this claim, I affiliate myself with the writings of scholars such as Philippe Pignarre and Isabelle Stengers (2011), Stefano Harney and Fred Moten (2013) and Anna Lowenhaupt Tsing (2015), all of whom provide an 'organic' critique of the university in their respective work. My ambitions are reflective of the historical contentions of ecocriticism at large – a subdiscipline that has always been at odds with academic professionalisation and its mandarin culture. For more on the resistant qualities of ecocriticism, see Dana Phillips (2003) and Gabriel Egan (2006).
2 For more on transversality, see Chapter 3 in this book (pp. 113–16).
3 There are many texts that deal with eco-activism and ecologically committed performances. Notable examples include Elisabeth Ellsworth and Jamie Kruse's curated volume *Making the Geologic Now: Response to Material Conditions of Contemporary Life* (Ellsworth and Kruse, 2012), T. J. Demos's *Decolonizing Nature: Contemporary Art and the Politics of Ecology* (Demos, 2016), Lisa Woynarski's *Ecodramaturgies: Theatre, Performance and Climate Change* (Woynarski, 2020), Bruno Latour and Peter Weibel's exhibition catalogue *Critical Zones: The Science and Politics of Landing on Earth* (Latour and Weibel, 2020) and Sarah Ann Standing's forthcoming publication *Provocation, Pathos, Protest: Eco-Activism as Performance*.
4 Raunig's book draws on the ideas of Guattari, but he does not appear to be as interested as I am in Guattari's insistence that the 'effects' generated by the artwork can never be predicated or policed nor that they emerge from its paradoxical autonomy.
5 For alternative models of theatre's relationship with activist politics, see Joe Kelleher (2009) and Alan Read (2008). A very different position is taken by Shannon Jackson who looks at how contemporary visual art and performance practices do 'social work' in more tangible and immediate ways, thus supporting the construction of new worlds in the very midst of this one (Jackson, 2011).
6 I stress the word 'indubitable', since activism contends that the action is intentional, something volitional or goal-directed, even in the case of clownish or absurdist guerilla performances that are concerned to disrupt the status quo without pointing to any definitive end. The irony, here, though, is that the aim

of performance can still be read and thus its incongruity easily interpreted, understood and thus negated.
7 One of the many difficulties in 'declaring' an emergency resides in the suggestion that the situation can be remedied and thus resolved without necessarily changing the structures that produced it. Another problem is that an act of declaration can be assumed to be *precisely that* – a mere speech act accomplishing nothing, a 'non-performative', to reference Sara Ahmed's critique of anti-racist performativity in what she calls 'institutional speech acts' (Ahmed, 2006: 104).
8 In the chapter on 'Natural Beauty', Adorno proposes a veritable ecological dialectic that ecocriticism would do well to remember: 'With human means art wants to realize the language of what is not human. The pure expression of artworks, freed from every thing-like interference, even from everything so-called natural, converges with nature' (Adorno, 1997: 100).
9 'Getting a taste', is very different from 'having taste' in the eighteenth-century understanding of the term. Whereas the latter is predetermined and already prescribed, the former is provisional, and contingent. Taste is something to be negotiated and invented; it involves becoming curious about things that lie beyond self – bodies, events, landscapes, animals, peoples.
10 This citation is taken from the title of Cage's *Diary: How to Improve the World (You Will Only Make Matters Worse)* (Cage, 1965).
11 The Capitalocene is a term that links the birth of climate change to capitalism. The Plantationocene plays a similar historical role but links climate change to Columbus's conquest of the Americas, the North Atlantic slave trade and the creation of the plantations in the Caribbean and Eastern seaboard of the United States of America. For more on these terms, see Introduction, Chapter 6 and the Glossary.
12 It is common to speak of the Anthropocene in terms of extinction. However, I wonder, after Sven Lindqvist's *Exterminate All the Brutes* (Lindqvist, 1992), if extermination is not a better word. Stengers's use of barbarism certainly underscores this point.
13 Apart from a handful of Guattarian scholars, *Chaosmosis* is a text that has been largely overlooked in favour of the shorter, more distilled *The Three Ecologies* (1989). I hope that my parsing of some of the ideas in *Chaosmosis* in Chapter 3 of this book encourages a return to Guattari's more rigorous, earlier account of ecosophy (see pp. 113–17). I would also like to thank my great friend David Williams for pointing out the similarities between Guattari's project and those of the Canadian poet Tim Lilburn in *Living in the World: As If It Were Home* (Lilburn, 2019).
14 Christel Stalpaert offers the term 'side-kick' to describe how the non-militant commitment of the artist Maria Lucia Cruz Correia relates to the dominant practices of green activist art (Stalpaert, 2018: 50). However, I prefer 'friendly enemy', in the respect to which the term maintains a greater sense of difference, a disjunctive relationality in which friend and enemy will always be opposed *and* yet always joined. As well as borrowing directly from Derrida, my use of the

phrase is influenced by Jean Genet's complex relationship with both the Black Panthers and Palestinians, as described in *Un captif amoureux* (1986).
15 Derrida's suspensive, impossible notion of friendship is directly opposed to the friend–enemy binary that is inherent to Carl Schmitt's notion of politics and which Bruno Latour, problematically, remains committed to in the seventh lecture of *Facing Gaia: Eight Lectures on the New Climatic Regime* (Latour, 2017: 220–54).
16 In Greek, *polemos* translates as war and warlike.
17 I take the image of ruins from Bill Readings's *The University in Ruins* (Readings, 1997).

Acknowledgements

No book is ever an individual product. It is a plethora, a dialogue, an ecology.

To honour that ecology, I would like to thank the following people for their intelligence, kindness and generosity.

To my readers: Melanie Lavery, Clare Finburgh-Delijani, Ralph Yarrow, David Williams, Alan Read, Joe Kelleher, David Archibald, Dimitris Eleftheriotis, Michael Bachmann, Liz Tomlin, Tim Barker, Eirini Nedelkopoulou, Graham Eatough, David Martin-Jones.

Thanks are also due to colleagues and students who listened to me raving on over the years. Lee Hassall, Simon Whitehead, Jonathan White, Harry Wilson, Cara Berger, Sarah Hopfinger, Isla Cowan, Laura Bissell, Misha Twitchin, Nik Wakefield, Christophe Triau, Vicky Angelaki, Cristina Delgado Garcia, Kristof van Baarle, Kris Verdonck, Graeme Miller, Rosa Casado, Mike Brookes, Minty Donald and Nick Millar, Maaike Bleeker, Christian Biet, Baz Kershaw, Lara Stevens, Peter Eckersall, Baptiste Buob, the Dodescaden, Nathalie Cau, Dominic Patterson, Ian Goode.

It would be remiss not to thank Matthew Frost and Paul Clarke from Manchester University Press for encouraging the project, the University of Glasgow for granting a semester of research leave and my colleagues in Theatre Studies for taking up the slack when I was absent from teaching and the increasingly relentless 'admin' we all face.

The sad Covid days in which this book was written were tough and I would like to thank the Tuesday night fireside group for their patience and tolerance: Martin, Jayne and David. A huge thanks, too, to Frank and Carl for showing me the way and offering hope and serenity in trying times. I took all the help that was offered.

As I was finalising the manuscript for this book in the lost summer of 2024 in Scotland, our dog, Mali, fell ill and died. Mali accompanied me on many 'a line of flight' through city, river and field – although her great love was the shoreline and the sea. I would like to thank her for the inspiration, fun and companionship. She is much missed by all of us; many tears have been shed.

The last thanks is to Melanie, as ever, for the talking, walking, reading, copy editing, creativity, hope and fortitude. Nothing would get done without you, and the debt will always remain unpaid. With love.

Introduction:
an idea for a theatre ecology

An unexpected project

I did not expect to write this book, the first of three loosely connected volumes investigating theatre's relationship with the earth. In 2018, when I started on this project, I had envisaged a conventional eighty-thousand word study that would explore, through a series of six tightly argued case studies, the different ways in which a select group of 'post-Artaudian' theatres affirmed the earth through a displacement of the human.[1] However, as I embarked on the work, it became obvious that something else was needed: a preliminary study that would theorise, in detail, the ecological potential inherent to the theatrical medium, introducing a new set of concepts for future critical and creative work and establishing a new genealogy for ecological performance. For while there is an ever-growing number of publications in the emergent 'field' of eco-theatre, many of which apply concepts from environmental philosophy and ecocriticism to performance, it is hard to find critics that are willing to reflect on the ecology immanent to the theatre event itself. Even more difficult to identify are scholars who look to gamble, as I do throughout these three volumes, on the possibility that theatre's ecological potential may be found in a terrestrial power that runs through the very heart of theatre like some unstoppable current, magnetising all involved, troubling their separateness.

Purposively, then, *An Idea for a Theatre Ecology* – the first book in the series – does not unpack the ecological meaning of performances, plays and performance lectures that deal with environmentally recognisable themes to do with climate change, extinction rates, oil production and colonial extraction. On the contrary, it looks to unfold an alternative theory of theatre ecology that reflects on methods, theories, histories and practices. The aim is to recalibrate what ecology and theatre are considered to be once they are approached as being implicated in and with each other, from the very beginning.

An Idea for a Theatre Ecology unfolds its argument patiently, taking the reader through its different stages and turns, step-by-step. In the first step, I argue for the necessity of a new perspective on theatre ecology by critiquing extant methods of doing ecology and environment in Theatre and Performance Studies; in the second, I provide a conceptual framework for understanding what the word 'ecology' in theatre ecology actually stands for by reading it with and against ecologically relevant models of posthumanist thought; in the third, I advance an alternative lexicon for unpacking theatre's immanent ecology; in the fourth, I construct a hidden genealogy for theatre ecology by returning to the earliest texts of western theatre theory and recasting the significance of the word 'theatricality'; in the fifth I disclose an extant example of ecological theatre, predicated on Antonin Artaud's notion of cruelty; and, in the sixth, I propose a new method of eco-performance analysis. Taken together, these steps underline the scope and scale of this monograph, its attempt to provide Theatre and Performance Studies with an immanent ecology that, on the one hand, will contribute to the work of thinkers and practitioners within those disciplines and, on the other hand, clarify the specific contribution that theatre can make to the Environmental Humanities, in general.

The idea I am promoting is only *an* idea, an indefinite proposition. In keeping with the thinking of the Martinican theorist and poet Édouard Glissant, I have no interest in establishing a totalitarian root. Such an absolutist agenda leads only to violence and coercion, a stultifying monologue for a deadly monoculture. Instead, I conceive of the idea operating in terms of what Glissant calls 'an open totality', a genuine invitation for others to reject or repurpose it, as they see fit (Glissant, 1997: 71).[2] Ultimately, all I have tried to do is to the reposition the lens through which theatre looks at ecology. The benefit of that repositioning responds to the hidden ecology that all theatre possesses by the simple fact that it asks audiences to assemble together at a particular time and place and to undergo a corporeal event that offers no real reason for its existence. I will return to the specific terms that concern this book shortly but for the moment I want to introduce the idea of theatre ecology that runs through the entire project from Volume 1 to Volume 3.

Premise

Theatre ecology, as I present it, is a kind of *dé-lire*, a non-human force that cannot be read (*lire*) or grasped. It overflows the understanding that would reduce it to an object to be represented or deciphered. To channel that flow, there is no need for theatre makers to deal with ecological matters in any explicit fashion; and neither is there a requirement, as some have proposed, to leave the 'theatre house' in order to make immediate contact with 'Nature' (whatever and wherever that paradoxical signifier is supposedly located). For if, as I contend, the theatre house is already haunted on the inside by what Gilles Deleuze and Félix Guattari term the 'monochrome infinite of the universe', then what *really* counts is to promote new practices – critical, creative and clinical ones – that are able to affirm the potential of that 'monochrome infinity' through an affirmation of theatre itself (Deleuze and Guattari, 1994: 180). Theatre has no need to import an ecology, because it already is ecological on account of its strange ambivalence, the doubleness that troubles any 'body', be that human and/or non-human, that has assented to come forward and show itself on a stage. In that appearing something else is sensed: a power of life as it alights and bifurcates, going elsewhere, a totality that splits.

In what I call theatre ecology (written purposively without a hyphen), theatre's taking (of) place is a strange and discomfiting event that shows the performer to be always in (at least) two places at the same time, both here and elsewhere; but also both human and nonhuman, material and immaterial, a creature – a body – that has assented to a process of mutation and metamorphosis that nothing can stop. When Richard Schechner made the celebrated comment that the actor (Laurence Olivier) playing 'Hamlet is both not Hamlet and *not not* Hamlet' (Schechner, 1985: 123), he was talking about human performance in the context of cultural rituals, the sense in which the actor inhabits a liminal space. By contrast, when I talk about the actor being haunted by what it is not, I am referring to what Gilbert Simondon terms the 'pre-individual', an energy flux that makes passage, like a hurricane, through the cosmos itself, and which all organisms – organic and inorganic – are hosts for (Simondon, 2020: 6).[3] Elizabeth Grosz offers insight into that haunting by citing the idea of useless production:

> Art and nature, art in nature, share a common structure: that of excessive and useless production – production for its own sake, production for the sake of profusion and differentiation. Art takes what it needs – the excess of colors, forms, materials – from the earth to produce its own excesses, sensations with a life of their own, sensation as 'nonorganic life'. (Grosz, 2008: 9)

In response to Grosz's terrestrial aesthetic of pre-individual forces, the ecological imperative of theatre, as I see it, is to make this excessive power palpable; to produce affective experiences able to tap and express what the dance scholar Erin Manning terms an 'insurgency of life', a surplus of being (Manning, 2016: 5).

This 'insurgency', this ecology, transforms the spatial parameters of theatre. Henceforth, the most appropriate metaphor is no longer architecture but meteorology. Theatre is a 'weather system', a dynamic vortex immersing spectators in its non-linear flight through time and space:

> Weather patterns are everywhere present in our everyday. Some of those presentnesses are artful, and some are not. Those that are artful are ones that make felt the intensity of material-forces. Not all art is relegated to the human realm, and not all art is artful. The artful makes felt the art of time, the event-time of the threshold, of the weather pattern. The artful is more than human. It is ecological at heart, multiple, serial. (Manning, 2016: 85)

Confronted with this inorganic ecology, the Greek etymology that has, for so long, posited theatre as a place for looking (*theatron*) needs to be recalibrated. For theatre, as Manning's writing on dance intimates, is a *sensorium*, a milieu for attentive looking, for taking the temper of the times, being on the lookout for the invisible, some impalpable contamination.[4] To 'watch theatre' is to be touched through the retina, to feel oneself corporealised (Fensham, 2009: 11–23). In theatre's construction of 'spacetimematter' (Barad, 2017: 103), performers and spectators are exposed to each other, bound together in clouds of affect that complicate the claims of those critics who demand to leave the stage for more 'authentic' connections with 'Nature'.

As these metaphors allow one to intuit, theatre's ecology is both terrestrial and cosmic. It uses bodies to get beyond bodies, to drench us in the im/material, in the ungraspable affects of life as it changes and evolves. Theatre ecology implicates, touches and transforms. It exists in terms of what Glissant calls 'un *écho-monde*', a reverberation in which the earth feedbacks into and rumbles through human-made discourses, images and sounds (Glissant, 1997: 208).[5] As *écho-monde*, theatre is immanently ecological, a noisy, turbulent medium where bodies find themselves plugged into a process of individuation in spite of themselves, a molecular flux that 'precedes the genesis of the individual' (Simondon, 2020: 48). To express that terrestrial flux, to convey those ecological frequencies, theatre is required to resist its all too human commitment to narrative and character. 'NOTHING … WILL HAVE TAKEN PLACE OTHER THAN THE PLACE' (Mallarmé, 2006: 178–9; original case), Stéphane Mallarmé says. And theatre is at its most profoundly ecological when it heeds his admonition of Aristotelian drama's obsession with narrated human behaviour, this 'vacuous action' (Mallarmé, 2006: 179).

For what Mallarmé appears to intuit is that to appear on a stage emptied of narrative is to participate in a force field, to be buffeted and billowed by terrestrial rhythms that are impossible to control. There is simply no choice in the matter. One always lives exposed. Such is life; such is theatre ecology. Theatre is an event that is always on its way, that prefers rhythm, cadence and speed to the anticipations and vicarious plotting of dramatic narration. Nothing, I believe, is as troubling as a body on a stage doing nothing. Its excessive presence is a violence, an affront to thought.⁶ For the temporary form that this body assumes, its apparent solidity, is always under attack by those same, pre-individual forces that have produced it in the very first place and have never stopped running through it. Theatre ecology is not about 'Nature' and nor is it *just* about being in relation. It is about attuning oneself to a dynamic force, an earthly power, a kind of storm.

Confronted with this storm, this stream of sensation, the performance analyst is stopped in their tracks, their surgical project butchered.⁷ In the presence of the *dé-lire* I touched on at pp. 2–3, the work of analysis can longer be content to engage in hermeneutics, to decipher the hidden meanings and structures of a text or performance by cutting it up into manageable pieces to make sense of. Rather, its responsibility is to stay resolutely on the surface of the performance, to undergo its ripple, to be affected by its pull. New questions arise from this illiterate ecology: where does the performance take me? What connections and practices does it produce? And, crucially, what are its politics? For, to borrow from Glissant (who in turn borrows from the poet Derek Walcott), to engage in and with ecological thinking is to open oneself to both matter and history, to the absoluteness of relation, to the swell of '*la totalité-monde*' (Glissant, 1997: 91). It is also to testify to the corporeal and poetic ways in which subjectivity is obligated to a planet-in-common, to a collective earth that is composed of discontinuous fragments that resonate and reverberate with each other, sometimes productively, sometimes harmfully. In Glissant's ecology, opacity triumphs over transparency every time. To get a taste for the opaque is to get a taste for life itself, to engage in a passionate adventure on earth, one that European imperialism looks to shut down, to immiserate.

Félix Guattari, too, stresses the need for ecology to escape its associations with sustainability and science. There is nothing reassuring about ecology, for Guattari – and it is for this reason that it must be expanded and championed as something dynamic and expansive, a life force:⁸

> Ecology must stop being associated with the image of a small nature-loving minority or with qualified specialists. Ecology in my sense questions the whole of subjectivity and capitalist power formations whose sweeping progress cannot be guaranteed to continue as it has for the past decade. (Guattari, 2000: 35)

To theorise and practise a theatre ecology worthy of the ecological thinking of Glissant and Guattari is to explain how ecology in the theatre takes shape from the inside, 'affording' becomings through the multiplicity of the relations and affective exchanges it puts into circulation. What I am arguing for, then, is a theatre ecology that offers itself as a politics, an aesthetics and a pharmacy; a nexus of thinking, feeling and creating that would contest extant ways of being a subject, a scholar, a spectator. If some are uncomfortable with this idea of ecology, believing that it says nothing about the environment, then so be it. The evidence so far suggests that art that 'talks' about ecology has achieved little. The more pressing task, as I see it, is to change direction, to think otherwise. The ecology that interests me is a variant on Timothy Morton's 'ecology without nature' (Morton, 2007), a fundamental ecology, an ecology that produces attachments to a kinetic earth. An earth that is unfinished, unpredictable, devoid of any reason or purpose.

The entirety of this study of theatre ecology unfolds over three volumes. Volume 1, *An Idea for a Theatre Ecology*, is largely theoretical, a metastudy that reflects on methods, concepts, models, histories and practices of performance analysis; Volume 2, *Theatre and Landscape: Taste, Ecology and Politics*, focuses on how theatre and performance can produce a planetary politics and ethics; and Volume 3, *Theatre as Clinic: Constellations of the Anthropocene(s)*, speculates on the clinical aspects of theatre ecology, using specific case studies to argue for how contemporary theatre can cultivate invigorating passions – ones that would rehabilitate and galvanise, restoring sensations and feelings that have been stolen from us by capitalist imperialism. I deliberately write Anthropocenes in the plural for the title of Volume 3 in order to highlight its contested and conflicted 'identity', the sense in which the Anthropocene is not an unmarked signifier for a universal process, but a terrestrial regime that is historically and politically situated, related to what Marxist and feminist new materialist critics such as Jason W. Moore (2013, 2016), Donna Haraway (2015, 2016, 2019) and Anna Lowenhaupt Tsing (2015, 2017) designate, respectively, as the Capitalocene and/or Plantationocene.[9] In such a context, therapeutics has little do with what is commonly thought of as theatre therapy. Rather, it wagers on the political-clinical quality of aesthetic experience itself, the immanent ways in which performance can 'heal' by sending subjectivity on journeys – sensate lines of flight on the lookout for alliances with other creatures, organisms, cultures, epistemologies, peoples.

In tandem with Guattari's ideas, these flights are ecological in the expanded way that I use the term: that is because they allow for the coming-into-being of creative modes of existence that have shrugged off the sadness and panic, the narcolepsy of integrated world capitalism's destructive desire for ever greater profit margins – what Philippe Pignarre and Isabelle Stengers term 'capitalist sorcery' (Pignarre and Stengers, 2011). At its best, by allowing audiences to take flight, theatre ecology can allow us to 'recover our belief in the world' and, as a consequence, to reconstruct what Guattari, the clinical ecologist, terms our 'entire mental ecology', a necessary component in a new aesthetics of planetary living:[10]

> Mental ecosophy will lead us to reinvent the relation of the subject to the body, to phantasm, to passage of time, to the mysteries of life and death. It will lead us to search for antidotes to telematic standardisation, the conformism of fashion, the manipulation of opinion by advertising, surveys. Its way of operating will be more like those of an artist, rather than of professional psychiatrists who are always haunted by an outmoded ideal of scientificity. (Guattari, 2000: 24)

Guattari leaves us in little doubt: ecology is not about saving 'Nature', it is about transforming the human subject in a radical fashion. That transformation has little truck with rhetoric or righteousness, the purpose is to produce a new ecological *habitus* – one predicated on adopting a creative attitude to life in all its forms. As I see it, theatre can act as a bridge for that transformation, a medium for facilitating passage from one world to another.

Idea

To turn now to the specific logic of this book, the idea that I am concerned with in *An Idea for a Theatre Ecology* is not, as it is for Plato, a timeless substance, an essence that holds good for every case and condition, and that can only ever be diminished by being actualised. Like any theatrical text, the idea calls out to be 'incarnated', given a body (Deleuze, 2004: 99). But that incarnation does not result in a simple mimesis, a relationship that unproblematically results in one-to-one correspondence, mere translation. For as everyone involved in theatre knows, actualisation changes origins and meanings, disclosing uncanny, anachronistic possibilities that were only ever latent or virtual within the initial script or score itself. The same process holds for the application of the theoretical idea, which, as Deleuze helpfully explains, is always undoing itself from within, questioning the very possibility of ever being used as a blueprint good for every situation: 'Beneath organization and specification, we discover nothing more than spatio-temporal dynamisms: that is to say, agitations of space, holes of time, pure syntheses of space, direction and rhythm' (Deleuze, 2004: 96).

In reference to several philosophers of the virtual – Henri Bergson, Gottfried Leibniz, Raymond Ruyer – Deleuze emphasises how the 'Idea is Dionysian' to its core. Beneath its Apollonian surface, there is an abyssal flux of impulses, processes and forces that are 'beautifully expressed by the image of murmuring, or the ocean, or water mill, or vanishing, even drunkenness' (Deleuze, 2004: 101). In these images without resemblance, these inebriated, chaotic rumblings, the idea 'reveals' itself, if indeed that is the right word, to be multiple and virtual. Something whose 'truth' or completion is perpetually deferred, only ever known by what it momentarily comes into being as, more of a gesture or provocation than a water-tight concept.

And so it is with the idea of theatre ecology that I am proposing in this text. In its attempt to remain faithful to this multiplicity, this book poses three fundamental questions: What is theatre? What is ecology? And how is the relation between them constituted? Yet to ask these apparently simple questions – questions that much extant work on theatre and ecology tends to leap over (see Chapter 1) – unleashes a plague of others: Where does theatre ecology emerge from? How does it fit with other ways of thinking ecology? What is unique about it? What types of theatre best express it? In what ways does it function? What does it afford? I look to interrogate these points in depth in the six chapters that follow. However, for now – and for the purposes of this Introduction – I want to reflect a little more on what I understand theatre ecology to be, why I leave it unhyphenated, and also to explain why it is so integral to the production of a renewed belief in the earth, the very thing that theatre ecology is so compelled to promote.

Theatre ecology

I take the term theatre ecology from Baz Kershaw's important 2007 text *Theatre Ecology: Environments and Performance Events*. For Kershaw, theatre ecology operates in both a restricted and a general sense. In its restricted sense, theatre ecology deals with performances that deliberately intervene into ecological and environmental debates, as, indeed, one might expect. In its more general usage, by contrast, theatre ecology relates to all those extraneous elements and systems that theatre is plugged into: audiences, economics, funding bodies, histories, performers, playhouses, archives etc. In this doubled coding – his ambitious and laudable attempt to embed theatre within a larger network of relations – Kershaw, in keeping with the work of Gregory Bateson, posits theatre ecology as a kind of cybernetics, a relationship of 'organism plus environment', a 'theatricalised ecology of mind', to pun on the title of Bateson's celebrated collection of essays from 1972 (Bateson, 1972: 455).[11] In plotting theatre's existence within a greater field of competing systems, Kershaw's aim is to provide a framework for 'beginning to evolve an ecohistoriography of theatre and performance for the ecological era' (Kershaw, 2007: 33). As far as I understand Kershaw, this would be a mode of historiography in which theatre is in constant exchange with multiple circuits of historical information that impact on it and which it feeds back into. Theatre, then, is not set apart from the social or the environmental, no matter how jealously or stupidly it clings to its autonomy. On the contrary, it occupies a niche within them, a part of a whole. But theatre's interconnectedness does not guarantee it a *de facto* ability to effect environmental change in any straightforwardly progressive manner. On the contrary, theatre ecology is a 'paradoxology', for Kershaw, on account of the fact that it does not – indeed, cannot – conform to the wishes and desire of an autonomous human agent (Kershaw, 2007: 23).[12] In Kershaw's cybernetic theory of performance, theatre, like everything else, is entangled from the beginning, simultaneously more and less than what it appears to be, always liable to outwit the desire of its authors, especially those expressly committed to proposing environmentalist solutions. To be a proper 'theatre ecologist', in Kershaw's terms, is to admit failure in advance, to come to terms with the primacy of the system, to relinquish control of the desire to impose one's own agenda on the world. When it comes to ecology, strength is a weakness, for Kershaw. Hence, the paradox of theatre ecology: the sense in which ecology is not about asserting one's willpower but accepting the distributed agency of the circuit.

While I certainly endorse Kershaw's embrace of relationality, contingency and paradox, my understanding of theatre ecology is somewhat different. Whereas Kershaw uses the term to talk of a network of relations – a general theatre ecology – I am more focused on what happens within the theatre event itself, with, that is, the aesthetic interplay between auditorium and stage.[13] Additionally, unlike Kershaw who is deeply suspicious of the eye's tendency to survey from a distance, to give the illusion of autonomy, I regard the eye as always embodied, an organ that is driven mad by sensation, *un dé-lirant*. To use a constellation of terms that I will explain on pp. 18–21, I am attentive to how the 'ecologics' of the medium allow theatre to 'ecologise' spectators, reminding them that they are terrestrial subjects, always in process, never coincident with what they are. Differently from Kershaw, theatre ecology, as I practise it in this book, is not just about placing theatre within a larger system of material and im/material relations and/or abandoning the theatre building in order to place them in contact with the earth; it is more concerned with tracking an immanent process of undoing that happens when the spectator and performance meet in a specially designated site that is both autonomous and heteronomous, human and non-human, at the same time. Otherwise put: theatre's outside is already inside. It is not just a question of making connection between differentiated systems or 'niches'.

The lack of a diacritical mark – a hyphen or oblique – between nouns in my 'writing' of theatre ecology is intended to signify this impossible and always incomplete interplay. For, while theatre and ecology are obviously relational, this relationality never achieves a sense of stability or balance – a oneness. I have little interest in synthesising the terms to provide a compound noun that would designate a hermetic genre of ecotheatre that could somehow – and damagingly – deal with ecological issues alone.[14] Instead, I want to posit an oscillating, uneven relationship between the two, in which theatre and ecology are simultaneously brought together and yet also separated from each other, like the impossible model of friendship that I discussed in the *Polemos* section (see p. xii).

The feminist philosopher Karen Barad sees this disjunctive monadology as constituted by 'diffractive relationships' (Barad, 2007: 69–94), in which one term of the equation is haunted by all the others that it both includes and excludes, a 'cutting together-apart' (Barad, 2014: 168). Theatre ecology is like 'water in water' (Bataille, 1992: 19), a crease within a fold, a term that destabilises the two terms that compose it, a doubleness in permanent oscillation. One can get a 'feel' for this confluence by simply speaking the word 'theatre ecology' out loud and by experimenting with different stresses and rhythms of the articulation. This concrete exercise highlights the variation in the noun in a manner that is both immediate and disconcerting, for it is ultimately impossible to separate theatre and ecology in any absolute sense. In one utterance, theatre is emphasised so that ecology becomes a theme or topic of/for theatre; and, in the other, ecology comes to prominence in such a way that theatre is an iteration of a larger ecology. But if to pronounce the word theatre ecology is to articulate a folded relationship in which the two words that compose it – theatre and ecology – exist as part of a disjunctive unity, it is nevertheless important to understand what each term in that heterogeneous synthesis refers to.

Theatre and theatricality

I use the word theatre in this book in a general and normative manner to describe a representational form in which bodies show themselves to other bodies on a stage or surface – a *plateau* – that has been specially designated for that purpose. The theatricality of theatre is found in the ambivalence of that operation, in the fact that a body on stage is always double, both itself and not itself, disjointed. While theatricality is part of theatre's ontology – at least in the West – not every form of theatre embraces theatricality in the same way. In fact, in western theatre, since Aristotle's *Poetics*, there has been a concerted effort to repress or occlude theatricality in order to produce what Hannah Worthen and Michel Corvin (see Chapters 4 and 5) define as a proto-humanist theatre founded on a reciprocal contract of suspended disbelief whereby the performer is taken as a vicarious representative of some fixed, substantive humanity – a surrogate for the spectator. Throughout most of its history in the West, the theatrical quality of theatre is perceived as troublesome, a shameful necessity, something to repress since, as I pointed out on p. 3, it shows the actor to be always more and less than they appear to be, a creature of artifice and metamorphosis – uncanny, maybe even monstrous, an unstable surrogate, on the verge of mutation.

The theatre that interests me is a theatre that makes a virtue of this uncanniness, this *semblance*. As the German theatre scholars Max Herrmann and Georg Fuchs put it somewhat tautologously more than a century ago, this is a 'theatrical theatre', a mode of embodied showing in which neither narrative nor the suspension of disbelief is integral to the ecological success of the performance.[15] This emphasis on theatrical theatre, a theatre that transmits sensations by drawing attention to corporeal gestures, explains the primary role played by the mid-twentieth-century theatre practitioner Antonin Artaud in my project. For of all theatre makers, Artaud is the one who realised that theatre is ecological on account of how its innate theatricality – what he called 'cruelty' – suspends all anthropocentric claims for identity, self-presence, exceptionalism (Artaud, 1958: 84–8). Artaudian cruelty is traumatic for Enlightenment notions of humanism that insist on ideas of human exceptionalism, the capacity to order and know the laws of 'Nature'. Ontologically and epistemologically, cruelty shows human subjects becoming undone by cosmic energies, victims of a terrestrial violence or 'plague' that they are powerless to withstand or reject (Artaud, 1958: 15–32). As I discuss in Chapter 5, Artaud's rejection of representation, his desire to transform the stage into a dynamic milieu in which actors and audiences encounter the bifurcations and diffractions of matter, sets a new precedent for theatre; one in which theatre emancipates itself from the comforting confines of its anthropocentric frame by accepting, finally, the non-human forces within it. In Artaud's thinking, theatricality is a cosmic earthquake, exposing all involved in it to 'suffering', in the affective and temporal senses of that word. For him, the doubleness of theatricality, its troubling of identity, is constitutive of life itself.

Ecology

As ought to be evident by now, the meaning of ecology in this book diverges both from its nineteenth-century origins in natural science and also from the way in which most theatre and performance thinkers have approached it as an analogue for 'Nature' and environment. As well as designating a biological milieu, ecology, for me, describes a type of agitation, an intensity, a theatrical fissuring in the very substance of being. Such an idea of ecology not only troubles notions of conservation and sustainability that have been so prevalent in environmental thinking; it also problematises the frugality that ecology has long championed.[16] As Philippe Lynes points out, ecology signifies an abundance of life, a surplus that cannot be counted or distributed out, a bond that goes beyond self-interest (Lynes, 2018: 102): 'The promise of this bond', he continues, 'is itself the promise of the earth, an ethical-ontological promise to *let life live on* and *let the earth be the earth*' (Lynes, 2018: 104; original italics).

In its promissory guise, ecology demands a kind of humility, an acceptance of weakness and incapacity. It opens up an earth that is not a place to appropriate or subject to some economic law of the hearth, as its semantic root in the Greek *oikologia* (economics) suggests. Rather ecology, to cite the environmental philosopher Michael Marder, is an 'event' (Marder, 2018: 142),[17] the expression of a creative planet, a geological form that has gone through multiple transitions, and which never stops transforming. In this context, Bronislaw Szerszynski's reworking of the Situationist International's notion of drift as planetary phenomenon offers a new agentic take on eco-aesthetics, a lithic theatre:

> And drift is old – a primordial power of the Earth. In the warm, young, abiotic Earth, before the 'zoic' aeons when life became dominant, matter's self-organizing powers could operate without constraint; and they innovated. The innovations of the drift were profound for the Earth. Each of these innovations has been rediscovered, again and again, by other non-drifting beings; but it is in drift that these innovations come together most powerfully and consequentially for our planet. (Szerszynski, 2018: 138–9)

To be on a drifting earth is to realise that life is precarious and contingent, a happening that came about through a brief and fortuitous period of recent climatic stability – what geologists call the 'Holocene', an era that started approximately eleven thousand years ago and runs to the present. Nevertheless the capricious indifference of the earth does not mean that humans are free to become inhuman, capitalist predators; and neither are they entitled to use the earth as mere resource or 'standing-reserve [*Bestand*]', to enlist a term from the phenomenologist Martin Heidegger (1977: 17). For to exist on the earth, to echo Lynes's 'aneconomic' notion of the bond, is to be interpellated by a non-human voice that one can neither place nor know, a silent frequency (Lynes, 2018: 106).[18]

As Jean-François Lyotard explains in his short but compelling essay '*Oikos*' (1989), the emission of that frequency reminds us of an unpayable debt, an obligation that inheres in matter itself, an anamnesis of an imperative that no memory can fathom or hold because it transcends the meaning-making structures of the *anthropos*. Anticipating the environmental ethics and politics of some posthumanist and new materialist thinkers, ecology is a saying 'yes' to life, a willingness to invent on a planet that belongs to no one and which embraces transience, uncertainty and incompleteness as creative catalysts – im/material forces that theatre is obliged to endorse because they form the basis of theatricality. In this context, what Elizabeth Grosz says about cosmic becoming below, holds good for theatre, too:

> At their most consistent and unchanging, beings are nevertheless points of convergence for an infinity of relations that ensure the entire system of things, the universe, is always changing, becoming. Any stability or foundation is itself relative to the instability and interactivity that marks the broader and further-reaching environment, the infinite connections between all things and all processes […]. This is an ethics without norms, without prescriptions […]. (Grosz, 2017: 260–1)

From this perspective, to affirm ecology is to affirm theatre and vice versa. In both, the imperative, always, is to learn how to 'de-create' oneself in a transient and incomplete spacetime, and to approach that de-creation as a moment of alliance and mixity with the immanent creativity of a cosmic earth. As with all earthly things, virtuality is immanent to theatrical experience. Any experienced spectator is aware that things are partial, that the actual – the performance one is attending – could be and will be different; that there is always another way of doing things, a possibility not yet taken, a staging that is already here but still to come. More than any other art form, theatre, as Samuel Weber has shown, is a medium that ungrounds, an exercise in de-phasing and deferral, a virtuality machine haunted by ghosts from the future (Weber, 2004: 1–30).[19] That machine does not work by representing the world but by making it teeter: more of a de-formance than a performance.

Affirming the earth

When I talk of the earth in these pages, I am not referring to a globe (too spherical), to a world (too human), or even to a planet when that is understood as a holistic system; rather I am concerned with something that is human and non-human at the same time, a dynamic, incomplete aggregate of sorts that humans are made from and return to, and yet which always transcends them. As I approach it, the earth is both material and immaterial, a place of storms *and* stars, localities *and* globalities, politics *and* peoples, technologies *and* plate tectonics. This earth is 'dynamic', unpredictable, and volatile, a planet disrupted by strange frequencies from within its own darkest recesses and open to cosmic forces that buffet it from the outside (Clark, 2011: 14). It is also a political entity, a figure for what Gayatari Spivak calls 'planetarity', an untranslatable word which demands careful negotiation when used (Spivak, 2014: 290). The imperative always is to avoid completion and transparency, to allow the earth to be earth, to acknowledge the provisional.[20]

Much of what has been sketched above aligns closely with Gilles Deleuze's and Félix Guattari's playful but influential notion of a 'geology of morals' (Deleuze and Guattari, 1987: 39–74).[21] For them, the earth is not, as it is for Martin Heidegger,[22] a source of authenticity, the home of being that one should learn to dwell in properly by constructing a house (*Bauen*) or by standing in a clearing in a wood (*Lichtung*).[23] Rather, the earth is cosmic, a multi-scalar planet affected by the pull of the moon, the rotations of the sun, weather systems, meteor strikes. The earth is not whole, it is constantly transforming:

> [T]he earth constantly carries out a movement of deterritorialisation on the spot, by which it goes beyond any territory: it is deterritorialising and deterritorialised. It merges with the movement of those who have had to leave their territory *en masse,* with crayfish that set off walking in file at the bottom of the water, with pilgrims or knights who ride a celestial line of flight. (Deleuze and Guattari, 1994: 85)

In the earth's deterritorialised performance, humans do not complete the earth by 'unconcealing its concealedness' in poetic speech. They are part *of* it, moved *by* it, and swept up *in* it. The terrestrial artist is someone who captures rhythms, transforming the artwork into a terrestrial intervention. In a key passage from their final text together, *What Is Philosophy?*, Deleuze and Guattari note how:

> We are not in the world; we become with the world; we become by contemplating it. Everything is vision, becoming. We become universes. Becoming animal, plant, molecular, becoming zero [...]. This is true of all the arts: what strange becomings unleash music across its melodic landscapes and its rhythmic characters, as Messiaen says, by combining the molecular and the cosmic, stars, atoms, birds in the same being of sensation? What terror haunts Van Gògh's head, caught in a becoming sunflower? (Deleuze and Guattari, 1994: 169–70)

Ecological being is nomadic being, for Deleuze and Guattari, always on the move. On Deleuze's and Guattari's earth, there is no destiny to recover, no origin to repair. There are only lines of flight, provisional and contingent openings without end or guarantee. Where the Heideggerian peasant returns home at the end of a weary day with the mud of the field on their shoes and the smell of burnt potatoes in their hair,[24] the nomad remains perpetually restless, a sunflower head, a deterritorialised ecologist:

> [T]he earth as ardent, eccentric, or intense focal point is outside the territory and exists only in the movement of D [deterritorialisation]. More than that, the earth, the glacial, is Deterritorialization par excellence: that is why it belongs to the Cosmos, and presents itself as the material through which human beings tap Cosmic forces. We could say that the earth, as deterritorialized, is itself the strict correlate of D. To the point that D can be called the creator of the earth – of a new land, a universe, not just a reterritorialization. (Deleuze and Guattari, 1987: 509)

Against all those that continue to posit art as a uniquely human activity, Deleuze's and Guattari's writings constantly refuse to oppose the cosmic earth to culture. On the contrary, the earth, for them, is the very source of culture, the thing that drives art, the force that artists have to feel, work with and transmit. The 'task of all art', they say, is to compose with terrestrial sensation and materials and so 'raise them to the height of the "earth song"' (Deleuze and Guattari, 1994: 176). Distinct from Heidegger, this song does not require the human to express it; it is an animal song, a stone song, a song that is sung by *any* 'being of sensation':

> [T]he being of sensation is not the flesh but the compound of nonhuman forces of the cosmos, of man's non-human becomings, and of the ambiguous house that exchanges and adjusts them, makes them whirl like winds. Flesh is only the developer, which disappears in what it develops: the compound of sensation. (Deleuze and Guattari, 1994: 183)

This notion of the 'ambiguous house' has far-reaching consequences for the idea and practice of theatre ecology in this project. For a cosmic earth can never be represented or narrated; it is a genesis not a thing:

> [B]ecoming is neither an imitation nor an experienced sympathy, nor even an imaginary identification. It is not resemblance […]. This something can be specified only as a sensation. It is a zone of indetermination, of indiscernibility, as if things, beasts and persons […] endlessly reach that point that immediately precedes their natural differentiation. (Deleuze and Guattari, 1994: 173)

To consider ecology as 'a zone in which we no longer know […] what is animal, vegetable, mineral or human in us', is to renounce interpretative reading, even if that reading is oblique (Deleuze and Guattari, 1994: 173–4). The more valuable work is to write out, as best one can, the indiscernible becomings that theatre can provoke:

> Contemplating is creating, the mystery of passive creation, sensation. Sensation fills out the plane of composition and is filled with itself by filling itself with what it contemplates: it is enjoyment and self-enjoyment. It is a subject, or rather an inject. Plotinus defined all things as contemplations, not only people and animals but plants, the earth, and rocks. (Deleuze and Guattari, 1994: 212)

There is no desire to separate history from ontology in what Deleuze and Guattari name 'the plane of composition' (Deleuze and Guattari, 1994: 192). Rather, a new type of politics – an earthly politics – is necessitated by this disturbance of boundaries, the enjoyment produced through 'the mystery of passive creation'. As Deleuze and Guattari insist again and again, the production of a new earth is unthinkable without the coming into being of a new people: 'In this submersion, it seems that there is extracted from chaos the shadow of a people to come in the form that art, but also philosophy and science, summon forth: mass-people, world-people, chaos people' (Deleuze and Guattari, 1994: 218). This 'chaos people' is an ecological people, a planetary people, an undercommons 'without the false image of enclosure' (Harney and Moten, 2013: 18). The North is not the guiding light for this undercommons, and neither do they look for an originary homeland or state to occupy and defend, as western politics so often encourages them to do. This new people know that no transcendent discourse situated beyond earth can save us. No god, no technology, no art, no philosophy. The imperative is to invent new forms of living together, experiments in becoming that transform suffering into possibilities, nihilism into life – what the contemporary decolonial thinker Achille Mbembe terms a new 'pharmacy' (Mbembe, 2019: 186). Theatre ecology, as I theorise it in this book, embodies that experiment. It looks to bastardise, to make improper, to trouble conventional ways of thinking about theatre and ecology in the name of a becoming that might restore us to health.

There is no doubt that my project has much in common with the aims of Indigenous ecologies that look to depart from the hegemony of western modernity, in particular the importance the latter attaches to humanism – the credo that places the human at the very centre of the world, the only animal that thinks and acts. In both instances, there is the same drive to celebrate an animated earth, to expand the notion of agency to plants, animals and minerals, and to interrogate the sacrosanct value that modernity places on the 'myth' of progress. However, despite these parallels, it is not my intention to engage in a comparative study with Indigenous thought in this text, no matter how valid or tempting that may be. Rather I prefer to follow the path of Glissant in the hope of finding another 'Europe' within Europe, one that would offer an alternative to white metaphysics, with its 'totalitarian' drive to make everything and everyone 'transparent' (Glissant,

1997: 11, 15).²⁵ By opening western thinking and practice to its own internal otherness, I look to discover the 'opacity' that, for Glissant, is the 'thing which protects Diversity' (Glissant, 1997: 62). Why? Because one can never figure or represent the excess that allows life to evolve and expand beyond what it is.

A note on style and shape

The writing in this monograph is mostly theoretical, drawing as it does on performance criticism, critical theory and posthumanist philosophy. However, this does not mean that its expression is overly technical. The book is meant for theatre and performance scholars as well as for students, practitioners, programmers and more general readers. So while there are certain conceptual terms that cannot be avoided – individuation, im/material, *oikos*, ecologising, intra-activity etc. – these are both discussed in the main text and further explained in a glossary. It is my intention that informed audiences can follow my argument, and I have done my utmost to explain my ideas in a language that is rigorous and nuanced without being arcane or overtly dense. I am, of course, aware of the paradox of that desire for clarity. As I have already intimated, the idea is something – a noise, an opacity – that resists capture in transparent discourse, something whose essence cannot be represented or completed.

As a way of holding that paradox in place, in acknowledging an impossible tension, I experiment with writing that would allow the affective stakes of this book, its 'mattering', to be communicated to the reader. The logic is to show how the non-human is always already at stake in our thinking, an excess that cannot be stilled, and which resists being domesticated in normative discourse. Consciously distancing myself from those critics in the Environmental Humanities that have recourse to autobiography in their attempts to practise a poetics of environmental experience – what Timothy Morton critiques as 'ecomimesis' (Morton, 2007: 29–79) – I express ecological affect at the more granular level of the sentence itself, in the use of syntax, choices of images and rhythm of the words.²⁶ I do this because one of the primary reasons for writing this book is to argue for a practice of theatre ecology that is attuned to the innate potential of theatre and performance to generate modes of perception that are not dependent on linguistic understanding or psychological identification, and which discomfort conventional modes of description.

As a way, then, of trying to communicate theatre's gestic, a-signifying mode of expression, I have attempted throughout this book to practise a mode of critical writing that taps preverbal 'percepts' and 'affects' which exceed and resist the intelligence of the *cogito*.[27] If we are serious about theatre as a site where one's ecological *habitus* can be re-created, it seems crucial to experiment with new forms of performative analysis. This is a response mode of analysis that tries to transmit some kind of pathos, to carry something else over and beyond its semantic meaning, to tap what happens in the realms of breath, blood, heartbeat. These are affects that break on me but which are not mine alone, belonging, as they do, to the neutral realm of haecceities, percepts that resist naming precisely because they are so full, lacking in nothing.

Outline

The book unfolds in the following fashion. Chapter 1: Contexts (questioning methods) starts by relating my idea of theatre ecology to congruent work already undertaken in Theatre and Performance Studies. Despite a relatively slow start in relation to the development of ecocriticism in the domain of Literary Studies, in the past decade or so Theatre Studies has witnessed a veritable explosion of new writings about ecology and environment. Faced with this proliferation – a mass of texts that one can never hope to map exhaustively – the key questions are no longer concerned with why Theatre Studies ought to engage with such things as extinction, pollution, environmental justice and climate change but rather *how* it does so, and *what* it hopes to achieve in the process. Posing these fundamental questions discloses hidden contradictions, unspoken assumptions and theoretical absences. To grasp better what is missing in the extant literature on ecology and environment, I focus on the disparate methods advanced for doing theatre ecology in what I consider to be the four leading texts in this area: Una Chaudhuri's *Staging Place: The Geography of Modern Drama* (1995), Bonnie Marranca's *Ecologies of Theater: Essays at the Century's Turning* (1996), Baz Kershaw's *Theatre Ecology: Environments and Performance Events* (2007) and Alan Read's *Theatre, Intimacy & Engagement: The Last Human Venue* (2008). I have chosen these publications for the way in which their differing approaches work together to compose a kind of deep structure in/for this field, offering a series of methodological underpinnings that most subsequent critical work in ecology and environment has tended to rely on, whether it is conscious of doing so or not. The irony is that while these works are routinely cited as exemplars and avatars, their 'affordances' have not been fully reflected upon, at least not in terms of their epistemological and ontological assumptions and commitments.

Chapter 2: Theory (theatricalising ecology) expands the study outwards by moving it beyond the discipline of Theatre Studies towards the

Environmental Humanities in general. The logic is to argue for the necessity of reading contemporary theatre in and through the disparate posthumanist models of ecology advanced by thinkers such as Vicky Kirby, Rosi Braidotti, Elizabeth Grosz, Claire Colebrook, Donna Haraway and Bruno Latour. However, while these influential posthumanist discourses play a key role in my thinking, I do not endorse their ideas, completely. Where posthumanist ecologies affirm, almost without exception, the power of narrative to produce new planetary imaginaries in a largely 'positive' manner, I am concerned to highlight the 'cruelty' of ecology, the sense in which matter is inhabited – and disturbed – by a theatrical impulse, a becoming other, that subjects it to a process of perpetual transformation and variation that audiences are compelled to suffer. Gilbert Simondon's notion of the 'pre-individual' plays a key role in this chapter, since it is the pre-individual that is integral to what he terms a 'theatre of individuation', a theatricalised understanding of life whereby every organism is shot through with an immanent but ungraspable power to become other than what is, that de-phases and disarticulates identity (Simondon, 2020:7).

Chapter 3: Model (a lexicon for theatre ecology) departs again from existing ways of working in both Theatre Studies and Environmental Humanities by constructing a posthumanist model that is simultaneously attuned to the theatricality of ecology and to the ecology of the theatre event itself. Particular attention is placed in this chapter on linking ecology to politics, aesthetics and existential individuation. In order to come to terms with these disparate qualities of theatre ecology, I place Karen Barad's thinking about the virtuality of matter in conjunction with the ideas of Lyotard and Guattari on aesthetic affect and ecological subjectification. In this way, I attempt to underline the 'doubleness' of my idea of theatre ecology, the sense in which the sensate impress of the aesthetic event produces a type of ecological transference that mirrors the relations, forces and bifurcations at play in matter itself. The intention of this modelling is not only to prepare the ground for the ecology that is found in theatrical theatres, but also to stress the need for a way of theorising that explicates the ecological work which theatre is already doing as opposed to making theatre conform to a monolithic idea imposed from without. The onus is on finding a language, not on erecting a grid or cage.

Chapter 4: Concept (ecologising theatre) moves from ecological theory back to theatre practice, while always attempting to keep the 'double fold' of theatre ecology in tension. It does so by revising Plato's anti-theatrical critique of theatre in the famous 'allegory of the cave' in *The Republic*. Where Plato sees theatricality as dangerous on account of its relationship with earthly becoming, I explain how these becomings are full of ecological potential, affects that derive from the vibrant presence of bodies that radiate with the same elemental intensity as the rocks in the cave. In the second part of the chapter, I use this posthumanist reading of Plato to argue for a new ecological conceptualisation of theatricality itself, one that diverges from traditional understandings that tend to equate it with the intentional acts of human agents alone. In my reading, which touches on the work of theatre scholars such as Erika Fischer-Lichte, Josette Féral and Samuel Weber, theatricality is ecologised and the theatre becomes a *plateau* for placing spectators in contact with the neutral forces of an agentic earth which are immanent to it, the very things that Plato diagnosed but recoiled from.

Chapter 5: History (Artaud's cruel ecology) advances an alternative, ecological history of contemporary western theatre that the French scholar Michel Corvin unintentionally and obliquely heralds in *L'Homme en trop: l'abhumanisme dans le théâtre contemporain* (Corvin, 2014), a text that is largely unknown in Anglophone Theatre Studies. In that publication, Corvin shows how the appearance of Symbolism in the late nineteenth century marked a break in western drama by deliberating looking to attune audiences to the unspeakable presence of the non-human on the stage – what he terms as '*abhumanisme*' (abhumanism). But where Corvin says nothing of ecology in his comprehensive history of contemporary European theatre – moving, as he does, through its dramatic, avant-garde and post-dramatic modalities – I situate an ecological shift in the mid-century work of Artaud who, as I have already hinted at, is the first performance maker in European theatre to draw a transversal line between earthly matters, processes of individuation and contemporary politics. As well as unsettling the boundaries between 'Nature' and 'Culture' in a manner that pre-empts, by almost a century, current new materialist models of thought at the level of content, Artaud also stressed the need to find new forms of 'theatrical theatre' that would allow the affective power of matter to transform spectators in and through the immanence of the performance event itself, in a manner akin to Barad's work on quantum agency. Such a positive take on Artaud's ecology necessitates a critical encounter with influential readings, including those offered by Jacques Derrida, that would either dehistoricise his ideas or place them, for understandable reasons, within a neocolonialist and authoritarian context. However, as I show, there is a different story to be told about Artaud, one that highlights his critique of western modernity's long history of environmental violence, its desire to colonise and control all of life.

Chapter 6: Analysis (becoming archipelagic) proposes a practice of ecoperformance analysis inspired by Lyotard's notion of the *gestus*. Where Bertolt Brecht sees the *gestus* as a making visible of signs that would articulate the *habitu*s of a character, allowing their historical lifeworlds and ideological belief systems to come to the surface and to be deciphered, Lyotard, by contrast, sees it as the way in which the materiality of the work (its colour, rhythms, textures etc.) is transmitted via a kind of affective impress. By insisting on the projectile power of the *gestus* to disrupt the intelligence and to target the prelinguistic forces of the body, Lyotard produces a mode of commentary alive to the ecological possibilities generated by Artaud's practice of theatricality. In line with Lyotard's understanding, and keeping Artaud's idea of cruelty firmly in mind, I contend that an ecological analysis should look to show how the subjectivity of the analyst is called into question by the somatics of performance in such a way that criticism's language of distance and objectivity is troubled. As a primary step in that more expressive direction, I offer, after Édouard Glissant, the figure of the archipelago as a terrestrial trope that brings together 'theatre' and 'ecology' in a geomorphic topology whereby each flows through the other and a whole series of human/non-human relations wash up together.

Notes

1 I use the phrase post-Artaudian in a non-normative manner to refer to twentieth- and twenty-first-century theatres that came both before and after Artaud. Artaud, for me, is an event in theatre history; his ideas illuminate the work of his predecessors as much as his successors, even when those practitioners make no explicit acknowledgement of his influence.
2 According to Glissant, an open totality is predicated on a restless relationality that never comes to a final conclusion. 'It is the boundless effort of the world to become realised in its totality, that is, to evade rest. [...] Relation relinks (relays), relates' (Glissant, 1997: 172–3).
3 For more on Simondon and the pre-individual, see Chapter 2 (pp. 97–101).
4 In French, to be on the lookout is translated by the verbal construct *donner sur*. In its major usage, *donner* translates as the act of giving something to someone. To watch over in the sense of *donner sur* implies, then, in the very construction of the word, at the very heart of the act, a type of looking that is a giving, a watching over, a gaze that implicates, in other words. One might also think here of Glissant's notion of *donner avec*, 'giving-on-and-with' (Glissant, 1997: 19, 212, n.5).
5 *Écho-monde* translates as a 'world-echo', or, better still, a world reverberation. It designates not only an attunement but a kind of musical accompaniment. In the extraordinary final essay 'Burning Beach' in *Poetics of Relation*, Glissant says that an '*echo-monde* is someone convulsed with chaos without realizing it' (Glissant, 1997: 208; original italics). The *écho-monde* is also a catalyst and conduit. Tellingly, and for reasons that will become apparent, Glissant posits Artaud as an *écho-monde* on account of his commitment to matter, to the earth (Glissant, 1997: 93).

6 This reading of performative bodies is indebted to Gilles Deleuze's 1981 text *Francis Bacon: The Logic of Sensation*. However, what Deleuze does not stress enough in that excellent text is that Bacon's canvases are effectively theatrical stages. In them, the figure is always raised up, separated from and made visible as a kind of solitary spectacle. This aspect of Bacon's work was evident in the 2019–20 retrospective of his work in the Georges Pompidou Centre in Paris.

7 Patrice Pavis says that any act of analysis is surgical in nature because it is predicated on cutting a performance up, creating what he calls 'a butchered effect' (Pavis, 2003: 8). This phrase resonates with the choice that Deleuze and Guattari make between 'dwelling as a poet or living as an assassin' (Deleuze and Guattari, 1987: 345). Perhaps, it is time to extend that choice to ecocriticism, too.

8 In early 2021 in the middle of a global pandemic caused by the outbreak of the Covid-19 virus, Guattari's expansive sense of ecology, first published in French in 1989, seems more vitally relevant now than ever.

9 Haraway initially used the term Plantationocene in a paper in 2015, but her usage is dependent on an unacknowledged but longer history of ecological and political work done on the plantation by Black Studies scholars such as Sylvia Wynter (1971), Édouard Glissant (1997) and Katherine McKittrick (2013). Moore first introduced the 'Capitalocene' *in Capitalism in the Web of Life: Ecology and the Accumulation of Capital* (Moore, 2013) and developed it further in the edited collection *Anthropocene or Capitalocene: Nature, History and the Crises of Capitalism* (Moore, 2016). He defines it in the following way: 'The Capitalocene does not stand for capitalism as an economic and social system. It is not a radical inflection of Green Arithmetic. Rather the Capitalocene signifies capitalism as a way of organizing nature – as a multispecies, situated, capitalist world ecology' (Moore, 2016: 6).

10 The beautiful phrase recovering 'our belief in the world' comes from Gilles Deleuze (1995: 176).

11 The full title of Bateson's book is *Steps to an Ecology of Mind: Collected Essays in Anthropology, Psychiatry, Evolution, and Epistemology*.

12 Since no message is without noise, no single actant or agent in the ecosystem – neither director, actor, spectator, nor critic – is able to predict, with any accuracy, the outcome of the meanings it purports to produce. Different contexts or milieux will vary that information and one is always liable to be undone by unforeseen consequences due to the multiple components involved in any ecological utterance or interplay. Paradoxically, the more one recognises one's interdependency, the greater one's freedom to play, to experiment with alternative strategies that avoid the contradictory interventions made by self-proclaimed environmental artists.

13 Reflecting on the title of his book, Kershaw insists that it possesses 'deliberate ambivalence', since 'it references both an object or objects of investigation, *theatre* ecology as in (*human* ecology), and the processes of theatre's unavoidable ecological engagement and/or disengagement regarding the environment, theatre *ecology*' (Kershaw, 2007: 10; original italics). He also claims, a

few pages later that, if pushed, 'theatre ecology is the way theatres behave as ecosystems. Or: The ecologies of theatre are the investigations of theatre ecosystems' (Kershaw, 2007: 16). One of the unhelpful consequences of this expanded or cybernetic way of thinking about theatre as an ecology is that it can refer to theatrical processes and research projects that have nothing environmental or earthly about them at all. While I am all for positing thinking, as Timothy Morton also does (Morton, 2010a: 1), as an ecological operation in itself, something connected and entangled, it is too easy for the 'performative society', as Kershaw, after Jon McKenzie (2001), puts it, to invest in this epistemology, without, however, being concerned with the ethics and politics of ecology at all. The very opposite, in fact. Ecological thinking, in Kershaw's sense, can be an excellent method for running a corporation or business that, in its 'ecological' commitment to sustainability, may well be perfectly willing to get rid of everything and everyone that stands in its way. No matter how expansive, theatre ecology should actively resist becoming an alibi for capitalist survival. For more on this point, see Chapter 2 in this book, pp. 84–91.

14 This strikes me as a classic example of an ecologically 'bad idea', in which 'a basic error propagates itself' and so produces a narrow, inflexible epistemology that can produce 'insanity' (Bateson, 1972: 492). Theatre ecology does not deal with a specific type of theatre, as if it were a subdivision in some housing project. It questions the very act of dividing altogether.
15 For more on the history and theory of 'theatrical theatre' in German Theatre Studies, see Erika Fischer-Lichte (2008: 32–6).
16 For more on the problematics of frugality, see Allan Stoekl's classic text *Bataille's Peak: Energy, Religion and Postsustainability* (Stoekl, 2007).
17 According to Marder, ecology is an event that 'announces itself when we feel not at home in our ecologically arranged homes' (Marder, 2018: 142).
18 'Aneconomics' is an economics that troubles the very idea of economics, expenditure that cannot be counted or reinvested to produce surplus value.
19 I am thinking here of scholars like Mike Pearson and Michael Shanks (2001), Marvin Carlson (2003) and Mary Luckhurst and Emilie Morin (2015) who are conscious that theatre is haunted by ghosts, *revenants*.
20 The untranslatability of Spivak's term 'planetary' is an attempt to acknowledge that the earth can never be signalled as a whole or globe. The earth is always elsewhere, undoing itself from the inside.
21 For more on Deleuze and Guattari on the earth, see Mark Bonta and John Protevi, *Deleuze and Geophilosophy* (Bonta and Protevi, 2003), Rudolph Gasché's *Geophilosophy: On Gilles Deleuze and Felix Guattari's What Is Philosophy* (Gasché, 2014) and Arun Saldanha and Hannah Stark's special volume of *Deleuze Studies*: 'New Earth: Deleuze and Guattari in the Anthropocene' (Saldanha and Stark, 2016). Other books worth consulting in this context include Bernd Herzogenrath's two edited volumes *[Un]likely Alliance: Thinking Environment(s) with Deleuze and Guattari* (Herzogenrath, 2008), *Deleuze/Guattari and Ecology* (Herzogenrath, 2009), and Jon Roffe and Hannah Stark's collection *Deleuze and the Non/Human* (Roffe and Stark, 2015).

22 In the crucial chapter 'Geophilosophy' in *What Is Philosophy?*, Deleuze and Guattari distance themselves from Heidegger's project of phenomenological deconstruction in no uncertain terms: 'He got the wrong people, earth, and blood', they claim (Deleuze and Guattari, 1994: 109). However, they also recognise that his notions of ungrounding, letting-be and critique of technology are 'close' to their own ideas of deterritorialisation (Deleuze and Guattari, 1994: 95). For an excellent study of Heidegger's and Deleuze's earthly politics, see Janae Scholtz's *The Invention of a People: Heidegger and Deleuze on Art and the Political* (2015).
23 I cite these terms from two of Heidegger's essays, 'Building Dwelling Thinking' (Heidegger, 1971b: 146) and 'The Origin of the Work of Art' (Heidegger, 1971a: 53).
24 I borrow this image from Heidegger's much-cited discussion of Van Gogh's painting *Old Shoes* in 'The Origin of the Work of Art' (Heidegger, 1971a: 33).
25 According to Glissant, 'the West has produced the variables to contradict its impressive trajectory every time. This is the way in which the West is not monolithic, and this is why it is surely necessary that it move toward entanglement' (Glissant, 1997: 191).
26 For Morton, ecomimesis is a curious and ultimately dubious activity. It uses writing to deny writing in order to show that 'Nature' is actually present on the page, existing in a space beyond the signifier (Morton, 2007: 31).
27 In *What Is Philosophy?*, Deleuze and Guattari define affects as 'no longer feelings or affections; they go beyond the strength of those who undergo them'; and percepts as 'no longer perceptions; they are independent of a state of those who experience them'. Affects and percepts belong to the non-human: 'they exist in the absence of man' (Deleuze and Guattari, 1994: 164).

1

Contexts (questioning methods)

Introduction

In an article on Niklas Luhmann's 1986 text *Ecological Communication*, the ecocritic Hannes Bergthaller advances the sobering suggestion that since the '"ecological crisis" has now persisted for well over a century and shows no signs of abating [...] we may wish to examine the concepts we often use unthinkingly, as if they referred to a world that was indifferent to how we communicate about it' (Bergthaller, 2018: 128). Although I do not endorse the full political implications of Bergthaller's notion of communication – following Luhmann, he seems to renounce a 'critical attitude to society' as naive and misguided – there is nevertheless much to be said for his concern to effect an epistemological shift in how humans first conceptualise and then represent their relations with 'Nature'.[1] For, as Bergthaller opines, in the midst of an ongoing 'ecological crisis' without apparent resolution, humanist concepts of conservation, sustainability and harmonious balance, once so integral to environmental thought, are no longer fit for purpose.[2] Rather what is needed, Bergthaller suggests, are more realistic, less sentimental discourses better able to engage with the 'permanent, irreversible change in the conditions of life' (Bergthaller, 2018: 18) that we are witnessing on a planetary scale in what Bruno Latour refers to as the 'new climatic regime' of the twenty-first century (Latour, 2018: 2).

In partial response to Bergthaller's provocation, this book proposes an alternative, theatrically focused model of ecocriticism for Theatre and Performance Studies. Unlike many critics, I do not assume that theatre is a good in itself, able to achieve a shift in ecological perception by its powers of representation alone; and neither am I interested in explaining the message of works whose axiomatic end is to make audiences aware of ecological dilemmas, along the line of the 'global challenges' model touted by so many UK universities today. My intervention is inspired by Antonin Artaud. In his extraordinary 1947 essay 'Van Gogh, the Man Suicided by Society', Artaud poses a basic but deadly serious question.

Addressing himself to an audience of critics and connoisseurs, Artaud asks 'What is the point of describing Van Gogh's painting?' (Artaud, 1976: 498). The same provocative naivety can be applied to theatre scholars interested in ecology and environment, too. What is the point of describing something that we already know, unpacking an environmentalist meaning that the performance is already telling us? Doesn't this merely alleviate spectators from the responsibility of co-creating with the work, being disturbed by it, in the same way that Artaud's identity was in his response to Van Gogh's canvases?

From the very outset of this project, then, there is an assumption that something is missing within extant ways of doing theatre ecology. To engage with that absence, I use this opening chapter to reflect on the primary means by which theatre scholars have sought to engage theatre with ecological debates over the past thirty years or so. There is an ecological logic embedded in this approach. For if ecology, at least in some understandings, is about organisms interacting in and with pre-existing environments, the fact that 'nothing comes from nothing', then ecological criticism is always a metacriticism of sorts, an approach that focuses on points of affiliation *and* bifurcation with extant contexts, assessing and *re-assessing* accepted modes of working, reflecting on what works and what does not. The inclusion of the hyphen in the spelling of 're-assessing' is crucial, here. For in the act of returning to canonical works, my intention is to disclose potentials within them that have yet to be realised, to recollect forward, so to speak.

Theatre criticism in the Anthropocene

As so many have said since 2000, the year when the term Anthropocene entered scientific discourse, to be a geological actant is to be beset with ambivalence.[3] On the one hand, there is the narcissistic knowledge that the human seems to have triumphed over 'Nature', to have left its mark everywhere, and, on the other, the traumatic realisation that our existence on the earth is precarious, contingent and, ultimately, transient. For to assume one's position as a terrestrial being is to realise that one is of the earth, and thus forced to share the same fate as all those creatures who lived and died on the planet as it went through its geological mutations: the Paleocene, Eocene, Pleistocene and Holocene. This latter awareness, as I explain in Chapter 4, troubles the anthropocentric logic of western modernity, predicated, as it is, on a model of humanist thinking, in which there is a destiny to fulfil, a truth to be discovered and an exceptional hero to follow.[4] When human beings recognise the full implications of what it means to live in the Anthropocene, a sense of disquiet is produced. Latour goes so far as to call it a type of 'madness', a collective derangement (Latour, 2015: 23). He is right to do so. Not only does this knowledge shatter the human's illusions of transcendental grandeur, it also calls its universalism into question. As so many feminist, posthumanist and postcolonial scholars have been arguing in the past decade or so, the nomenclature of the Anthropocene conceals its own violence. In the same way that not everyone is responsible in the same way, so not everyone undergoes the violence of the earth equally or at the same rate. As Kathryn Yusoff has pointed out, there have already been 'a billion black anthropocenes', historical instances where 'black and brown bodies have taken the burden of exposure to toxicities and to buffer the violence of the earth' (Yusoff, 2017: 11).

To think of time in the Anthropocene is to contest the *telos* of Anthropocenic temporality, the desire of scientists to cling to western modernity's investment in Aristotelian narratology with its schema of a beginning, middle and end – a denouement.[5] The Anthropocene shows that ending is coterminous with the beginning, and that the future is already at work in the past. What Aristotelian could ever hope to disentangle the vertiginous coursing of time whereby the huge spike in fossil-fuel consumption in the nineteenth and twentieth centuries was not only trading on death that happened over three billion years ago, but is simultaneously still burning up the future of the twenty-first century in the very act of doing so? Anthropocene knowledge is disruptive and temporally promiscuous, something that the Belgian philosopher of science Isabelle Stengers equates with 'the return of pharmacological memory' (Stengers, 2010: 35). This is a type of memory founded on the necessary poison or evil of the *pharmakon*, a figure that complicates any attempt to offer a total solution to an irresolvable problem, as so many ecomodernist and technofuturists look to do. The *pharmakon* shows reality to be doubled and duplicitous. No positive without a negative, no life without death, no one without the other. In proximity to the *pharmakon*, totality and immunity are futile passions. We are contaminated creatures, a condition that disrupts the transcendental logic and ethics of western philosophy.[6] As a consequence of the *pharmakon*'s uncanny capacity for generating oxymorons, tearing things 'together', the redemptive logic of binary thinking is foreclosed. Now the poison is the cure and vice versa. Sickness is necessary for health. One can never escape the *pharmakon*; we are condemned to recognise its presence, to listen to its call, to live compromised. This is what terrestrial being entails.

This return of the ambivalent figure of the *pharmakon*, along with the agonising, irresolvable dilemmas it poses, has crucial implications for theatre practice and scholarship that few have reflected upon with any kind of urgency. For if to live in the Anthropocene is to remember that triumph is also defeat, there seems to be little point in arguing for models of ecotheatre and ecocriticism that would champion the potential of theatre and performance to restore problematic notions of balance, living at right scale, in oneness. Harmony is foreclosed by pharmacological memory. We are born bifurcated. And neither is there much (if indeed anything) to be gained in supporting practices and theories that look to engage, critically, with phenomena such as climate change or ecocide, while neglecting to question the hidden assumptions (the intentionality) inherent in their own logics of representation and models of subjectivity. Despite their ecological claims to the contrary, the human subject often remains central in such practices, unfazed by its predicament, confident in its capacity to resolve matters. Things need to change, for practitioners, critics and publics alike. To exist in the 'so-called' Anthropocene is to accept a deeply unsettling regime of knowledge, one that interferes with what Erin Manning names 'the volition-intentionality-agency triad' (Manning, 2016: 6), so dear to the anthropocentric *habitus* of capitalist subjectivity, the self of the 'Moderns', that in Latour's thinking has made the disastrous move of trying to sever 'Nature' from 'Culture' (Latour, 2018: 16–17).

If we are to have a theatre ecology worthy of the doubleness of the *pharmakon*, and not one that reassures us with absolute solutions, then it is important to fashion critical approaches that are attuned to what they can *and* cannot do. Merely *applying* ecological ideas from other fields or assuming an *a priori* notion of performativity as a good in itself, does not really cut it. Something more destabilising and discomfiting is required for Theatre and Performance Studies. Increasingly, we need to assess the 'efficacy' of what we are professing, to subject it to a kind of radical appraisal, acknowledging not just what Nicole Seymour terms 'bad feelings' (Seymour, 2018), but also 'bad faith', to recycle an almost forgotten term from Jean-Paul Sartre (1989: 47–70). But where to start this work? Perhaps by asking a series of fundamental questions about methods? Which ones have come to the fore, and why? How are they applied? Are they robust enough to effect the epistemological shift they claim necessary for survival?

Questions such as these are, *undoubtedly*, more pressing for Theatre and Performance Studies today than for cognate areas such as Literary Studies and Environmental Philosophy. For where those disciplines have longstanding engagements with ecology and environment and so are already attuned to the need to interrogate the 'value' of ecocriticism (Clark, 2019), it is only recently that Theatre and Performance scholarship has turned its attention to the non-human world. Consequently, the imperative has been to catch up;[7] and, in the past decade or so, the overriding feature – the thing that has been championed – is not self-reflection or evaluative critique but simple production: the impulse to produce more, to make up for lost time.[8]

This imperative has proved successful – at least to an extent. To peruse the titles and content lists of many books and journals in Theatre and Performance Studies today is to find much evidence of work on objects, animals, insects, climate change, weather, wildness, water management systems, agrilogistics, dark ecology, sexology, pollens, allergens, environmental justice, cities, rivers, mountains. These analyses have been explored through a number of different lenses, including phenomenology, affect theory, gender theory, new materialism, queerness, critical race studies, postcolonialism, decolonisation, Indigeneity, anthropology, autoethnography and disability studies. There has also been a medium-specific concern, as one might expect, to address the *ecologics* immanent to the theatre event itself. In the main, this latter concern has concentrated on such things as the ecology of rehearsal processes, practice-based enquiries into the performativity of objects and matter, experiments with 'more-than-human' performance scores and, of course, the ecological footprint left by the energy demands that twenty-first-century theatre requires, either inside or outside of buildings. New terms and neologisms have been invented to articulate these practices: 'ecoscenography', the Anthrop(o)scene, ecospectating, green theatre and, most popular of all, 'ecodramaturgy', a type of theatre making that, as Wendy Arons and Theresa J. May concur, puts 'ecological reciprocity and community at the heart of its theatrical and thematic intent' (Arons and May, 2012: 4).[9]

Inevitably, perhaps, the enthusiastic drive by scholars, editors and research councils to validate the medium, to show what theatre is able to do, has meant that important preliminary questions have been forgotten. Little time, for instance, has been spent reflecting on the affordances of the medium and/or interrogating the methodological frame used. The conventional format has been to highlight the ecological problem at stake (say climate change or flooding); introduce a pre-existing ecological concept to address the problem; apply that concept to a performance or specific case study; and wrap up by making positive noises about theatre's environmental efficacy. This is not to say that such an approach lacks value, it is simply to acknowledge that its contribution is inevitably limited by the range and ambition of the project. And it is worthy of note – and again a sign of performance anxiety – that the majority of publications to date have been special editions of journals, collections of essays and stand-alone essays,

Happily, three notable exceptions appeared while I was writing this text: Julie Hudson's *The Environment on Stage: Scenery or Shapeshifter* (Hudson, 2020), Lisa Woynarski's *Ecodramaturgies: Theatre Performance and Climate Change* (Woynarski, 2020) and Theresa J. May's *Earth Matters on Stage: Ecology and Environment in American Theatre* (May, 2020). Hudson's wide-ranging study looks to account for how spectators produce ecological meaning from their encounters with lively and agentic matter on stage. Woynarski's book is equally diverse but her concern is with tracing a series of intersectional, decolonial and Indigenous ecologies as they play out in a number of different ecodramaturgical practices. And May, more drawn to playtexts, excavates a buried ecological history in US theatre. However, for all the care, rigour and thematic originality, the respective methodologies of these monographs nevertheless rely on orthodox methods of 'reading' performance in which the critical task is to unpack the ecological significance that a given mise-en-scène, performance intervention, eco-installation or site-specific production can produce. So while the perspectives adopted in these publications certainly widen our vision of what 'ecotheatre' might be, they do not necessarily question its legitimacy or interrogate its relevance. Indeed, with respect to content, they remain conventional, concerned, as they generally are, with making sense of performances that are relatively explicit in how they represent 'Nature' and engage with the environment. In all cases, the medium is still not seen as an ecological apparatus in and by itself – something that ecologises through its own material intervention in the world.[10]

A related critique can be made of their practices of performance analysis. In each of their texts, the function of analysis is expository, the tone of criticism enthusiastic, mostly serving to draw positive attention to how theatre and performance can contribute to a better, more progressive ecology. In doing so, however, they never put criticism 'in crisis' or problematise their own capacity to digest and unpack what the theatre pieces they explain are supposedly saying, despite the fact that Hudson is aware of how some productions can be 'ecoconsciously unconscious or unconsciously ecoconscious' (Hudson, 2020: 6). In these monographs, extant humanist methods of writing about ecology are perpetuated (the objectivity of description, analysis and judgement), even when the critic's identity is 'situated', as it is in Woynarski's case, or phenomenologically foregrounded as an 'I' that feels, as in Hudson's.

I do not intend these criticisms to be harsh or pointed. I have great admiration for these new additions to the field, and neither Hudson, Woynarski nor May is alone in their reluctance to stray too far from tried and tested methods of doing theatre ecology. As I argue below, their texts, like most of the ecocritical work in Theatre and Performance for that matter, remain, consciously and/or unconsciously, indebted to four major monographs: Una Chaudhuri's *Staging Place: The Geography of Modern Drama* (Chaudhuri, 1995), Bonnie Marranca's *Ecologies of Theater: Essays at the Century Turning* (Marranca, 1996), Baz Kershaw's *Theatre Ecology: Environments and Performance Events* (Kershaw, 2007) and Alan Read's *Theatre, Intimacy & Engagement: The Last Venue* (Read, 2008).

Though these publications have not all had the same degree of influence, each of them proposes an original approach to theatre ecology, supported by a rigorous thesis, distinctive history and particular methodology. Somewhat bizarrely, for all the citations these writers have gathered in recent years, especially Chaudhuri and Kershaw, little has been done to tease out the complexity and/or plausibility of their arguments.[11] Too often, their ideas have been diligently referenced, without any real attempt to understand the profound disagreements between them and/or to assess the ecological and political consequences of what those disagreements entail for theatre's ability to make a contribution to a progressive planetary ecology.

Likewise, there has been no attempt to think through why some methods have met with greater critical purchase than others, and if that emphasis is something that ought to be revised and recalibrated. In Marranca's case, for instance, the biocentric form of landscape theatre that she was the first to propose has been almost entirely obscured by Chaudhuri's and Fuchs's more empirically focused and historically inflected understanding of the term in their collection *Land/Scape/Theater* (Chaudhuri and Fuchs, 2002).[12] Similarly, it is only recently that Alan Read's rigorous ecological questioning of the human subject in *Theatre, Intimacy & Engagement: The Last Human Venue* has been discussed, and, even then, the full aesthetic-political potential of his disorientating notion of theatre as the 'last human venue' still remains to be teased out. In responding to these gaps and oversights, my aim is not to impose some correct, monolithic method, and neither is it simply critical (although that *is* part of the exercise). Instead, and to return to an idea mentioned in *Polemos*, the goal is to be productive (see p. xiii). I want to use theory to ask scholars, practitioners and students to rethink their habitual ways of doing theatre ecology, to get a better understanding of the specificity of theatre's environmental contribution. The stakes could hardly be higher. Any glance at the news or casual conversation about the weather is enough to prove that.

For those familiar with the work of critics mentioned (as well as with contemporary developments in this area of study), my choice of texts may seem curious, a too-limited distillation, perhaps. In the decades since these monographs appeared, the thinking of at least three of the authors – Chaudhuri, Kershaw and Read – has evolved in quite different directions and new topics and agents and actants of ecology have come to the fore in their research.[13] So why, one may ask, am I concerned to make this seemingly anachronistic, maybe even reductive move? The answer is that I wanted to get at something more basic, the obscured ground upon which the suppositions and premises of their ideas are founded. For while the content has certainly changed, with new subjects, interests and themes coming to the fore, the work of the four critics mentioned has remained consistent on the methodological plane, the level, that is, where one encounters not only the epistemological and ontological assumptions subtending the conceptual aspects of the argument but also what the argument is supposedly intended to achieve beyond the aesthetic realm – its praxis, in other words. In ecocriticism (as indeed in all criticism), methodologies adhere to and express a latent, but always palpable, politics, ethics and aesthetics. Before any actual agent enters the scene, they tell us what ecology is, how the artwork engages with it and where the critic or spectator should place their attention. In the language of the perceptual psychologist J. J. Gibson, methodologies allow us to understand what theatre – or rather a particular perspective on theatre – is able to *afford*.[14] Namely, the ability to gauge the specific types of thinking, perceiving and behaving that the theatrical habitat, like any sensory-material environment, gives to those people who are involved *with* and *in* it, a milieu of capacities and possibilities.

Differently, then, from most scholars who have been content either to affirm or to ignore the methods advanced in the pioneering studies of Chaudhuri, Marranca, Kershaw and Read, I want to return to them in detail, to tease out their theoretical commitments, and thus to gain some insight into their practical affordances. The ambition is not just to understand what the field of theatre ecology looks like in Theatre and Performance Studies today; it is to imagine what it may need to become, to theorise alternative practices, to call out for new methods – ones that not only may transform how theatre ecology is practised theoretically but also how it is made artistically. For, as I argue in this book, there is something decidedly theatrical about ecology and vice versa, a doubled relay that impacts on what ecotheatre is habitually imagined to be. The function of ecocriticism is changed, too, by this hypothesis, called upon to justify its mode of analysis and discursive style, to become the transformative thing it purports to be.

Staging Place: The Geography of Modern Drama

Una Chaudhuri's 1995 monograph *Staging Place: The Geography of Modern Drama* remains a pivotal work in theatre ecology. Its pioneering contribution is to have traced a complex thread of exile and belonging in modern drama from the late nineteenth century through to the 1990s.[15] Chaudhuri starts her ecological history by showing how naturalist and realist playwrights dealt, negatively and obliquely, with the environmental devastation of nineteenth-century industrialisation by presenting audiences with characters desperate to leave home, suffering from what she calls 'the geopathic idealisation of ill placement' (Chaudhuri, 1995: 20). Staged tragically as suicide (*Miss Julie*), ironically, in indecisive thinking (*The Cherry Orchard*) or positively in heroic departure (*A Doll's House*), the 'geopathology' that naturalism promotes, is, for Chaudhuri, an exercise in displacement, a symptomology (Chaudhuri, 1995: xi). For what is really at stake – and she stresses this very point in her conclusion to *Staging Place* – is a refusal to acknowledge, indeed to wilfully repress, the violence inflicted upon 'Nature' by western modernity.

In the second part of her argument, Chaudhuri reverses geometrical perspective and traces a trajectory of 'failed homecoming' in the work of Absurdist and late modernist playwrights such as Samuel Beckett, Harold Pinter, Tennessee Williams and Caryl Churchill. For these mid-century practitioners, writing in the shadow of Nagasaki and Hiroshima, the 'end' of empire and the great acceleration in petroleum production, the ability to flee is no longer an option. Instead, the only direction possible is a movement of retreat. This return to natality, however, does not end in restoration or repair. Natal being, for these playwrights, is ill-being, something to be endured melancholically, an experience of abandonment after the event, 'a victimage of location from which no heroism of departure can rescue them' (Chaudhuri, 1995: xiii).

Dissenting from standard existential and psychoanalytical readings of the Absurd, Chaudhuri demonstrates how it is the planet (and not just language) that is falling apart in the work of late modernist playwrights. To apply the words of Old Ekdal, the displaced and alienated woodsman from Ibsen's *The Wild Duck*, 'the forests' in Absurdist theatre 'have taken their revenge'.[16] And they have done so in such a way that people, plants and animals are ruined. The damage is everywhere: in how one thinks, feels, moves and relates. By shifting the point of attention from humans to non-humans, Chaudhuri is the first critic to allow us to see how, in the work of playwrights like Samuel Beckett, say, absurd being becomes a negative iteration of ecological being, offering dramatic insights into what the Marxist ecologist John Bellamy Foster terms a 'metabolic rift' in capitalist production, the fissuring that has done so much to create the 'sixth great extinction event' (Bellamy Foster, 2000: 155–63).[17]

The third and final part of Chaudhuri's argument is more hopeful.[18] It concentrates on how postmodernist practitioners in the US (Laurie Anderson, Suzi Lori-Parks, Tony Kuschner, Ping-Chong, Spalding Gray etc.) have accepted that there is no longer any proper place or home to escape and/or return to. For all its destruction and violence, globalisation has performed a kind of dialectical ruse, Chaudhuri maintains, by tying humans to each other as well as to the earth, thus allowing for a post-pathological sense of dwelling to be intuited in nascent form in the multicultural practices of the writers and communities in what she calls 'America'.

In this new platiality, the trope of American placelessness, according to Chaudhuri, is now recast as a 'redemptive figure', forming the basis for what the ecocritic Ursula Heise, in the late 2000s, would come to define as 'eco-cosmopolitanism' (Heise, 2008: 11) and what Bruno Latour has theorised as 'landing' – learning to live on the earth (Latour, 2018: 4):

> Through its alliance with the principles of progress and of homogeneity, the figure of America first signified a kind of ultimate placelessness, a guarantee of the absolute *un*meaning of place as a component of human experience. But the very success of this figuration – what one might call the hyperbole of American utopianism – proved to be its undoing. In the late twentieth century, the figure of America has begun to be required, increasingly to make good its utopian claims, and the principle of placelessness is confronted by the multivoiced demand for new *placements*. (Chaudhuri, 1995: 5; original italics)

In addition to endowing modern drama with a new history of dwelling, Chaudhuri makes telling insights into the ecological potential of postmodernism's experimentation with form. According to Chaudhuri, the early practitioners of modern drama initially tried to accommodate lost 'Nature' by representing it positively and transparently in the same way that nineteenth-century photographers and architects displayed it on acetate and in glass palaces and botanic gardens. The 'garden' that Old Ekdal built in the loft in Ibsen's *Wild Duck* is a key *topos* in this regard. Symbolically 'located directly behind a photographic studio' (Chaudhuri, 1995: 76), the garden looks to synthesise incompatibles, to merge the domestic with the wild, spectacle with reality. But this attempt to heal an ontological and epistemological wound – this felt sense of 'Nature's' loss – is futile, Chaudhuri concludes. 'The Ekdal loft' exists not only as 'a symbolic space but as a symptomatic one', a simulacrum which, in keeping with modernity's entire aesthetico-technical outlook, aggravates and 'distorts' extant divisions between 'nature and artifice' (Chaudhuri, 1995: 76).

By excluding what it purported to include, modern drama's early simulations did not act as a panacea or agent of repair. On the contrary, 'man-made nature', as Chaudhuri calls it, produced a feeling of 'uncanniness' that intensified the malaise, giving rise to a headlong flight from reality into hyperreality:

> The exemplary product of this collision is a structure that has become rather familiar to us all, a hundred years after Ibsen's play: the artificial environment. Perhaps our experience of this phenomenon may shed some retrospective light on Ibsen's prescient model of the paradox of a man-made nature. The main

feature of this paradox is that advances in mimetic technology (of which photography was only the beginning) produce strange worlds of irreducible *strangeness*. (Chaudhuri, 1995: 76; original italics)

Abutting on the arguments of historians and geographers such as Keith Thomas (1984) and William Cronon (1996), Chaudhuri alerts readers to the fact that 'Nature' on the naturalist stage is merely a cultural construct, an ideological signifier. A set of artificial protocols and mythological tropes that attempted to resolve contradictions by producing sentimental attachments to a world from which people and things have been forcibly expunged:

> The harsh reality of an aggressive capitalism underlies the sentimentalism of the Ekdal loft as much as it did that of the charming winter palaces of the time. In both capitalist exploitation – be it of forests or people – requires that nature be artificially reproduced, preserved, and displayed. The end product of this process is something with far reaching consequences in modern experience, especially in the modern theatre: it is the naturalization of the artificial. (Chaudhuri, 1995: 78)

Chaudhuri's argument effects a stunning shift in traditional aesthetic taxonomies that remains highly relevant to ecocriticism today. For by suggesting that naturalist drama allowed 'western man to stage himself as lord of the imperilled green world' (Chaudhuri, 1995: 77), she transforms naturalism into what it always was: a stylistics. In doing so, she is able to rehabilitate postmodernism, the cultural poetics that supposedly replaced the 'really real' with a simulated world of appearance. Conversely – and this is what makes Chaudhuri's book so theoretically rich as well as deeply ironic – postmodernism allows for a new engagement with the natural world by deconstructing the very values that environmentalism clings to and cherishes: the 'Natural', the 'immediate', the 'eternal'. Reversing the position of Fredric Jameson and harking back to Bergthaller's call for new concepts (see p. 25),[19] Chaudhuri argues that postmodernist performance produces a more 'truthful' understanding of 'Nature' – one that is paradoxically 'real', purged, as it now is, of naturalism's delusional belief that it could represent the world, make it knowable, render it present through signs.[20]

A special place is reserved in Chaudhuri's ecocritical revision of postmodernist theatre for Spalding Gray's solo piece *The Terrors of Pleasure: The House* (1986), a performance that, as she puts it in a companion essay in the 1994 special edition of *Theater*, deliberately addresses the environmental and spatial 'ruptures' inflicted by consumerist capitalism (Chaudhuri, 1994: 28). In the production, Gray's pastoral dream of finding Romantic solace and retreat in the Catskill Mountains in upstate New York is hilariously undone by the work of real 'Nature'. Beset with subsidence, dry rot, broken pipes and dead skunks, Gray's *oikos* – the Greek word for house – is subject to entropy, a condition illustrating the second law of thermodynamics. This is the law which states that every system or organism is involved in a random and chaotic process of disintegration over time. In showing how 'Nature' is not something to be domesticated through a spew of nostalgic images, but something with its own inhuman agency, Gray manages, Chaudhuri claims, to produce a viable 'ecotheatre', in which 'the rupture with nature so long, so anxiously (and often so surreptitiously) registered by the drama of this [the twentieth] century is brought into the dramatic spotlight' (Chaudhuri, 1995: 90).

Through Gray's postmodernist version of 'negative dialectics', a new relationship between theatre and environment starts to emerge, characterised, dramaturgically and scenographically, by a refusal of spectacle and new emphasis on frugality. In the performance Gray sits on the stage alone, accompanied only by a table, a microphone, and a glass of water. The 'poverty' of Gray's scenic apparatus is central to its ecological significance. Its simplicity disturbs the picturesque view of 'Nature' that both naturalism and the contemporary real estate industry equally depend on. In the process, Gray transforms theatre into a new 'kind' of home, an actual place to be inhabited for a specific time and in a special way, not exclusively through ownership, but through 'the direct sharing of experience' (Chaudhuri, 1995: 83).

Chaudhuri's analysis of Gray's work marks a generative tension in her thinking that warrants attention. Here the ecological potential of theatre is no longer simply equated with what it narrates or speaks of, as it is in the rest of Chaudhuri's *Staging Place*. Rather, it is found in the particular affordances of the medium itself: in the way in which theatre, when stripped back to spatio-temporal basics, exists as a collective dwelling space, where one can find temporary shelter from the homelessness caused by capitalism's desire to make a drama out of crisis. The very thing, perversely, that so many ecodramatists today still have recourse to in their concern to tell a good story.[21]

So when Chaudhuri proposes Gray's theatre as the 'model (as well as the theorisation) for an ecotheater', it is important to note that she is not simply talking of formal or dramaturgical issues – at least not in this instance (Chaudhuri, 1995: 90). Rather, she is thinking about theatre's spatial ontology. For her, theatre's insistence on taking place endows it with a possibility for ecological transformation that other modes of representation do not so easily possess. Gray's theatre is more than representation: it is an actual praxis, a concrete ecological intervention. It grounds spectators in the here and now.[22]

Gray's ecotheatre accomplishes this work of 'grounding' in two ways, Chaudhuri explains. On the one hand, his performance undoes naturalism's falsification of 'Nature' by refusing to indulge in elaborate fictional narratives that present themselves as 'real'; and, on the other, it proposes theatre as an empirical site where audiences can remake their world *together*. As Chauduri has it, Gray's model of ecotheatre roots actors and audience to the spot, transforming them into subjects of gravity, earthing them.

This insistence on gravity, in celebrating theatre's spatio-temporal irreducibility, is borne out in her analysis of Ping Chong's *Nuit Blanche: A Select View of Earthlings* (1981), a decentred, multi-media performance that reflects on select episodes of human history on earth in order to argue for a progressive multiculturalism – one predicated on a complex universalism of sameness *and* difference. Chong's play is acutely aware that there is a negative side to multiculturalism, a destructive monologue that would reduce everything to abstraction, the terrible logic of the self-same. In *Nuit Blanche*, the dangers of a totalitarian oneness are intimated from the very start of the performance via a slideshow sequence entitled 'Murmurs of the Earth'. The slideshow is based on the archive of photographs and sounds that were sent into space on the Voyager II mission in the hope of explaining what human existence is to alien life forms (should they ever be encountered).

Anticipating the philosopher Kelly Oliver's argument in her 2015 publication *Earth and World: Philosophy after the Apollo Missions*, in which the blue-marble view of the planet is deconstructed, Chaudhuri shows how the rest of *Nuit Blanche* ironises and critiques the series of 'high-tech moon shots' projected during the 'Murmurs of the Earth' sequence.[23] The difficulty with these images, the performance implies, is that they perpetuate the dangerous illusions of naturalism: namely, that one can adopt the view from nowhere, and grasp 'Nature' as it 'really is', a universal substance that everyone has access to in the same ways.

In opposition to this point of view, the ending of *Nuit Blanche* depicts a dying monster on a blood-red stage, above which a film is projected, once again, showing the image of a moon. But this moon is a very different moon from the one depicted in the 'Murmurs of the Earth' section. It is a moon obscured, seen through the branches of a tree. By ending on images of a sublunar moon, 'viewed from the perspective of earth-bound human beings', *Nuit Blanche* offers an alternative way of dwelling on the earth, Chaudhuri contends (Chaudhuri, 1995: 248). One that is necessarily and ineradicably rooted in the particularities of place: 'unlike the high-tech moon shots of the "Murmurs of the Earth" sequence, the moon that is seen here is "our" moon' (Chaudhuri, 1995: 248). That obscured, fragmented moon is necessarily ecological and political: the thing that allows for terrestrial difference to be respected. 'Sublunary existence', Chaudhuri proposes, 'binds us together and shows us exactly what we are in danger of losing; the "other view", the abstracted view from "outside", produces only distance, disconnection, destruction' (Chaudhuri, 1995: 248).

For Chaudhuri, the whole point of a progressive 'ecotheater', the 'poor one' that she most associates with Gray's work, is to contest this abstracted, technological view of the earth, which it can best do by revealing what naturalist theatre is so concerned to veil: partiality, opacity, contingency. Like the image of the earth taken from space that no terrestrial human could ever hope to see, let alone inhabit, naturalism, like today's ecomodernism, reduces the planet to an abstract globe, something that lacks density and weight, a groundless object to be controlled, measured and managed.

As a consequence of its ambitious scale, its desire to rethink the place of 'Nature' in modern drama, Chaudhuri's monograph opens up a generative reading of theatre ecology, in such a way that issues of content, form and medium-specificity need to be taken into account. So while her approach is certainly hermeneutic, concerned, as she always is, with reading for environmental meaning, Chaudhuri is able to transcend the limitations inherent in many 'green readings' of classic texts (Shakespeare, in particular, suffers from this approach). All too often, these readings overlook the material specificity of theatre's mode of being: its capacity to take place, to offer real spaces of ecological solace and invention. As I have demonstrated, and irrespective of its influence, there is a real sense that 'we' are still learning to read *Staging Place*, grappling with its complex tensions, its fine-grained and always dialectical ways of thinking about what a terrestrial, sublunary theatre entails. For what Chaudhuri teaches us is that theatre's platiality holds out the possibility for a different mode of dwelling, an *oikos* that is discovered in our willingness to attend to something that is happening in front of us, *all* of us, in the here and now, together.

Ecologies of Theater: Essays at the Century Turning

Published one year after *Staging Place*, Bonnie Marranca's *Ecologies of Theater: Essays at the Century Turning* is a collection of essays and reflections that asks what it means to make and experience theatre in 'a more worldly way' (Marranca, 1996: xiv). Narrative drama has a relatively minor role to play in Marranca's 'natural history of performance' (Marranca, 1996: xvi). Instead, her commentary is marked by a series of lyrical insights into the ecological connections and resonances between theatre, sound art, performance and horticulture. Yet notwithstanding this eclecticism, Marranca, no less than Chaudhuri, is concerned, at all times, to situate her thinking of ecology as a thinking of theatre, an art form that, for her, occupies a key position in contemporary culture, in so far as its *modus operandi* – the unmaking and remaking of identity – remains central to the concerns of a performative society, no matter its increasingly precarious and marginalised existence within that society. Marranca is adamant about her commitment:

> The question before us now is not so much how to make a living in the theatre, but how to make a life in the theatre. Where does theatre, with its obvious identity crisis, fit in the culture at this time of crisis in contemporary life? Even more to the point, what will theatre do now that life itself is experienced more and more according to a theatrical paradigm? What is it that will make being in the theatre important to us today? (Marranca, 1996: 287)

Marranca's response to these resonant questions is to unearth what Chaudhuri's symptomatic and fraught history of modern drama has to forget, if it is to make sense. Namely, the presence of an alternative, more spiritual and celebratory ecohistory of theatre rooted in avant-garde experimentation and centred on the work of Gertrude Stein, John Cage, Robert Wilson, Heiner Müller and Rachel Rosenthal:

> In this sense, the American avant-garde is tied historically to the spiritual dimension of European modernism in the beginnings of abstraction at the turn of the century and its early decades and to Buddhism in the postwar years. It was important to enquire into the relationship between mind, space, and spirit as a formal issue before I could find the way to an ecology. (Marranca, 1996: xiii)

As her final sentence intimates, theatre ecology, for Marranca, is a formal affair, primarily related to how avant-garde practitioners transform the stage into a living place, a new ecosystem.[24] In an essay on Heiner Müller, for instance, Marranca writes that his theatre is nothing other than 'the history of figures in a landscape' (Marranca, 1996: 78); and, in relation to Stein and Cage, she is quick to describe how their work, by 'reaching out to the natural world, to nature as process', is infused with a 'feeling of open air' (Marranca, 1996: 21).

In Marranca's vision of theatre as landscape, the spectator is encouraged to make conjunctive leaps between bodies, gestures and images that move in and through theatrical time and space like so many organisms within an environment.[25] Ecology, like theatre, is 'compositional', for Marranca, and the stage a milieu in which 'art is made to be more like life than the other way around' (Marranca, 1996: 21). In this move from representation to presentation, this insistence on the corporeal presence of theatrical bodies, new affirmative emotions are solicited. Where Chaudhuri sees rupture in modern drama, Marranca, by contrast, finds '*a climate very rich with joy*'

(Marranca, 1996: 31; original italics). And it is surely telling, in her history of avant-garde performance, that theatre should be so often compared to a landscape, garden or a field, a place to grow things in:

> In the same period [1930s], Stein began to define her notion of the play as a landscape in a radically original book of nature. This spatial conception of dramaturgy elaborates the new, modern sense of a dramatic field as performance space, with its multiple and simultaneous centres of focus and activity replacing the conventional nineteenth-century fixed setting of the drama. The effect is a kind of conceptual mapping, in which the activity of thought itself creates an experience. A more expressive understanding of this idea is the Roman sense of the mind as field, that is, site of cultivation. Now part of the common vocabulary of contemporary practice, the concept of performance space, which opposes the character demands and causality of setting, really begins with her theatre. (Marranca, 1996: 6–7)

To track this process of 'theatrical cultivation', Marranca dispenses, necessarily, with hermeneutics as a methodology. Her ecocritical method is to stay on the surface of things, to draw a tessellated map in which every organism, including humans, form part of what she calls, after Gregory Bateson, a larger 'ecology of mind' (Marranca, 1996: xviii). In Marranca's relational understanding of ecology – and here she anticipates Kershaw by more than a decade – theatre has its own internal and external ecology, on account of the fact that it is both an autonomous 'collective of texts, images, or sounds' and a heteronomous niche bound up with external 'production, reception, and funding, as well as institutions and interacting artistic communities' (Marranca, 1996: xiv).[26]

According to Marranca, theatre practice evinces a biocentric ethic when it gives up on Platonic and Aristotelian notions of mimesis as ratio and instead looks to establish relations between different signifying systems, each of which insists on its autonomy within a common, compositional field. Sounding very much like Bateson, Marranca posits Wilson's scenographic theatre as a conduit for producing an expanding ecological consciousness, a way of overcoming separation:

> Wilson's dramaturgy imagines a world in which it is possible to comprehend that a cultural system cannot be separated from an ecosystem, and that cultural change is impressed upon the landscape. His theatre acknowledges the place of nature in human consciousness. (Marranca, 1996: 48)

There is much to commend in Marranca's biocentric view of theatre ecology, not least the still highly original insight that the formal experimentations of avant-gardist and postmodernist performance encourage spectators to experiment with what Timothy Morton in 2010 would come to call 'ecological thought' – 'a practice and process of becoming fully aware of how beings are connected with other beings – animal, vegetable, or mineral' (Morton, 2010: 7). For that reason alone, her book stands in significant counterpoint to Chaudhuri's practice of theatre ecology. For Marranca, theatre is 'a more-than verbal medium', an instrument for expanding sense and democratising perception, something in which metaphysical concepts of appearance and reality, and ideas of inside and outside, no longer hold. In a passage on John Cage's sonic ecology, Marranca notes how:

> For Cage each sound was alive, a living organism. All species were welcome in this landscape but were left uncultivated. It is as if they were 'escapees' from the garden into the wild. *Let sounds be themselves.* Seen in this light, groups of sound can be comprehended as ecosystems, which in turn are linked to sociocultural systems. Whether animate or inanimate, they are all equal voices in the environment. (Marranca, 1996: 30; original italics)

Irrespective of the fact that her writings have yet to attain the same influence as those of Chaudhuri, Marranca's work merits serious ecocritical scrutiny. Not simply because she was the first to draw attention to a still neglected ecological heritage within modernist performance practices, but, more importantly, because she reconfigures what theatre ecology is and, thus, should do. What Marranca shows is that one can no longer oppose 'Nature' to 'Culture', theatre to the earth. Instead one is compelled to think them together, to overcome the separation that modernity is always desperate to construct between human and non-human worlds. Differently from Chaudhuri who, in spite of her interest in theatre as a spatial site, tends to alight on environmental themes, Marranca sees theatre ecology as a decidedly formal practice, a way of organising bodies within an open field. For what high modernist theatre highlights, Marranca contends, is the extent to which we are – all of us – organisms in biotic environments, living landscapes.

Fifteen years after the publication of *Ecologies of Theater*, Timothy Morton neatly summarises the thinking behind Marranca's theory of theatre ecology: 'Ecological art and the ecological-ness of all art, isn't just *about* something (trees, mountains, animals, pollution, and so forth). Ecological art *is* something or maybe it *does* something. Art is ecological insofar as it is made from materials and exists in the world' (Morton, 2010a: 11; original italics). Morton's reference to doing reveals the stakes of Marranca's work. It emphasises how, for Marranca, theatre ecology is a praxis that asks spectators to perform ecology in the very act of watching bodies in time and space. In keeping with the ideas of the Australian eco-feminist Val Plumwood whose 1993 book *Feminism and the Mastery of Nature* was published at roughly the same time as *Ecologies of Theater*, Marranca 'foregrounds' what so many ecocritics 'background' in their desire to get under the surface of things (Plumwood, 1993: 4): the irremediable presence of the earth in theatre, the sense in which it is – and always has been – a biocentric art form.[27]

Theatre Ecology: Environments and Performance Events

Written a decade after Chaudhuri's and Marranca's initial interventions and haunted by the refusal of neoliberal governments in both the Global North and South to confront the tangible accelerations of climate change, Baz Kershaw's *Theatre Ecology: Environments and Performance Events* (2007) adopts a more critical and provisional approach to theatre's environmental potential than his predecessors. Part of that criticality is due to the alternative theory of ecology that he adopts. Ecology, for Kershaw, is not about 'Nature' and 'Environment' *per se*. It concerns what it means to live connected, part of a system that one can never apparently transcend.

Good systems theorist that he is, Kershaw realises that, in a cybernetic network or circuit, our humanist ways of tackling a problem directly are undone by complexity and indeterminacy – the autopoesis of the system itself.[28] Strongly influenced by Bateson's work on the double bind[29] and Jon McKenzie's *Perform or Else: From Discipline to Performance* (2001),[30] Kershaw posits a 'paradoxology' of performance as something to embrace, a possible solution to a performative culture, addicted to action, hung up on willpower and intentionality. The aim is not to try to change ecological consciousness by working within the same anthropocentric paradigm that produced the problem in the first place. Rather one needs to approach the problem from the side, tempering the human being's attachment to agency and volition. In keeping with the logic of paradox, obliqueness, for Kershaw, is a strength, an indirect strategy for outwitting the pretensions of the ego, the pilot that tragically always wants to steer the ship, that recognises nothing beyond its own will to perform.[31]

Kershaw's 'paradoxology of performance' has important repercussions for how ecotheory is approached in Theatre and Performance Studies. Whereas Kershaw certainly welcomes Chaudhuri's and Marranca's contributions, he is nevertheless cognisant of how a key contradiction in their thinking compromises their arguments. For him, this centres on their reluctance to critique the engrained anthropocentricism of the medium itself. Conceived as a black box or Italianate stage, theatre, for Kershaw, operates in the same way as Ekdal's loft does in Chaudhuri's symptomological reading of *The Wild Duck*. That is to say, it is still imagined as an ocular apparatus, a site that posits 'Nature' as an image to be consumed, contemplated and ultimately mastered by a spectator who remains outside of the system in operation. By trying to solve the problem with the same tools that have created the problem, Chaudhuri and Marranca (and also Elinor Fuchs) continue to replay modernity's disastrous separation of subject and object – the ontological relation that, for Kershaw, is at the root of the ecological crisis:[32]

> Marranca, Fuchs and Chaudhuri are possibly all correct in identifying postmodern performance as a strong area for such revisions [the nature/culture binary], because its reflexivity may most crucially challenge the dualisms of modernism which have fuelled the ecological crisis, particularly those between mind and body, analysis and creativity, thought and action, spectator and participant, culture and nature. But clearly to the extent that postmodern performance in *theatre* depends on a sustained separation between performer and spectator, however ironically framed, then it risks replaying the tropes – of landscape, of pastoral, of wilderness, for example – that may reinforce the source of the environmental nightmare in the human. (Kershaw, 2007: 316; original italics)

In *Theatre Ecology*, Kershaw seeks to lessen the harmful 'double bind' inherent in Chaudhuri's and Marranca's models of ecoperformance by advancing three alternative suggestions.[33] First, he urges artists and scholars to experiment with site-specific and immersive modes of performance that would avoid the distance – the spectacle – of footlights.[34] Against the ocularcentrism of the *theatron*, Kershaw puts the onus on an 'eco-phenomenology of participation', in taking part in what he conceives to be an authentic 'environmental event':[35]

> Immersive performance events which are articulated directly to what's left of the 'natural world', unlike theatre, may have the capacity to collapse that disjuncture, to suture more fully human 'nature' with nature's 'nature'. They might achieve this in ways that will not reverse the first decisive act that led to civilisation [the clearing of forests], but which could lead humans to a fuller appreciation of how they are a wholly integral part of the earth's environment, acting *in* it rather than *on* it. (Kershaw, 2007: 318; original italics; citation modified)

Kershaw's second method is to dispense with all pretension of aesthetic autonomy and to posit theatre, as a systems theorist would, as an informational unit within a greater ecology of mind, an instance of 'an organism + environment'. One of the ways that Kershaw effects this move is to consider theatre as a whole (its mode of production, its relationship to other forms of spectacle, its audience numbers) as an ecology. Thus, for Kershaw, and on this particular point he draws close to Marranca's model, as I have previously noted (see p. 40), theatre ecology is not simply equated with an ecological theatre that would only represent (if it ever could) environmental issues. Conversely, it figures theatre as a particular mode of communication constituted by a larger circuit of information that it simultaneously affects and is affected by. Theatre is a type of feedback, in other words:[36]

> Hence 'theatre ecology' is the subject of 'ecologies of theater' and likewise 'performance ecology' of 'ecologies of performance'. The main implication in all these definitions is that theatre and performance in all their manifestations always involve the interrelational interdependence of 'organisms-in-environments'. Or to reformulate that last phrase from a more radical ecological perspective as proposed by 'deep ecologist' Arnae Naess, they constitute a 'relational total-field' in which everything is interdependent and cannot always easily be assigned to clear distinctions, say as between 'organism' and 'environment'. (Kershaw, 2007: 16)

In keeping with these necessary circumlocutions and complicated neologisms, theatre ecology is more than mere metaphor in Kershaw's usage. Rather – and this brings me to his third contribution – it is an ecology, an actual organism that exists in a homologous or isomorphic relation with other natural systems:

> The ecology of theatre is fabulously more than a metaphorical enterprise by how they [ecologists] have understood the fine balance of energies that can make or break the life of events. Hence theatre ecology is as much a matter of living exchange between organisms and environments as the ecology of plants or of ants, say. It operates much to the same principles as they do, and for purposes that are much more similar than is conventionally imagined. In other words, theatrical performance is not a system that is different in kind from ecological systems, though of course like them it has its own peculiar characteristics. (Kershaw, 2007: 24)

Throughout the book, it is notable how Kershaw borrows terms from the natural sciences – particularly from biology, chemistry and experimental physics – to disturb the binary oppositions between theatre and Nature, and arts and science that have pertained to western humanist thought. His homologies include positing theatre as an ecotone, 'a place where two or more ecologies meet and intermingle' (Kershaw, 2007: 19); reading certain forms of activist theatre as 'black holes' that sometimes get sucked into the force field they seek to combat (Kershaw, 2007: 255–74);[37] and approaching avant-garde performance as a 'free radical' that can produce a shift in the current system by acting as 'energy out of place' (Kershaw, 2007: 295):

> The transformation of energy in and between materials is the process that expresses vitality, and without appropriate vitality performance is not likely to bring about change. Hence a convincing general account of the powers of performance to induce change will rest on cogent figuring of how energy circulates in theatre and performance ecologies. (Kershaw, 2007: 275)

In tandem with the ecophenomenological logic that motivates his ideas on site-specificity and immersion, Kershaw's notion of metabolic exchange – an energetics – offers a generative way of gauging the difference that theatre ecology can make. Here, theatre is important not because it promises some actual encounter with 'real nature', but for the ways it produces 'transversal exchanges' between different ecologies and ecosystems, a term that Kershaw takes from Guattari. To think transversally is to affirm both unity and difference at the same time, to cut across supposedly separate domains, to assemble heterogenous parts together in a dynamic, open-ended totality (Kershaw, 2007: 261–2).[38] Since performance is a transversal agent – a machine for making connections, for plugging in to other circuits and systems – it is able to have a direct ecological impact out of all proportion with its apparently limited aesthetic remit. Indeed, for Kershaw, it has the potential to undermine the toxic models of performativity that are so omnipresent in the twenty-first century, to produce a different metabolism from the addictive ones that exhaust all resources (Kershaw, 207: 318):[39]

> As a now compulsive desire for performance is the motor of the addiction, fresh views on how it might operate on the world are required. Key to this, I think, becomes understanding how performance is an integral part of global ecology and ecosystems, how particular performances operate ecologically as ecosystems, and how humans might arrive at a better appreciation of performance processes in Earth's ecosystems through ecology. (Kershaw, 2007: 14)

It is precisely because performance can have real ecological benefits, for Kershaw, that it is important to approach it with accuracy, reflecting rigorously on habitual ways of seeing so as to evade double binds.[40] After Kershaw, ecotheatre is no longer about representing nature or producing environmental messages in conventional activist ways. Its primary task is to establish a new epistemology, a different take on knowledge. What is required ultimately is an investment in paradox, the confidence to affirm the distributed agency of the system or milieu, a willingness to believe that positive results happen in ways that bypass the intentionality of the will.

Some indication of how a paradoxology of performance might function concretely is found in Kershaw's analysis of Earth First!'s 'Smokey the Bear' intervention at a Forest Service children's birthday party in Covallis Oregon, in 1985. Concerned to highlight the disturbing fact that the Forest Service 'are ten times more likely to be the source of fires than anything else' (Kershaw, 2007: 269), Mike Roselle of the environmental group Earth First! adopted a very different approach to activism. Differently from those committed performances that wear their critique on their sleeves, so to speak, leaving no doubt about the seriousness of the situation, Roselle infiltrated the party in the apparently harmless, apolitical guise of a Smokey the Bear mascot. Once inside the party, Roselle started distributing leaflets, informing spectators of the environmentally negligent activities of the logging companies officially endorsed by the Fire Service. Predictably, his acts soon drew the ire of a Park Ranger who started to fight with Smokey Bear/Roselle. This produced an absurdist but decidedly unedifying situation in which the Forest Service lost its carefully crafted image as a purveyor of the environmental good and disclosed itself as it really was: an organisation committed to economically unsustainable practices. In this way, 'farce', according to Kershaw, becomes an oblique method for unveiling the 'green washing' that the Fire Service birthday party event was intended to occlude (Kershaw, 2007: 269). By witnessing an actual intervention, spectators are asked to think for themselves, to piece together the environmental incongruities of the situation at hand. In doing so, Roselle and Earth First! managed to avoid activism's blind spot: the self-defeating paradox in which the activist work, by telling you what to think, all too often contradicts its own environmental premise by investing in the same epistemology of the society it wants to change. This is an epistemology which contends that human reason alone is able to solve its own problems, without taking into account its position within a larger system of communication that necessarily compromises reason's autonomy.

To demonstrate his argument further, Kershaw shows how the spectacular dramaturgy at work in Greenpeace's Brent Spar protest in the North Atlantic in 1995 merely repeated the heroic images and thought processes of Shell UK, the multinational oil company that its protest was apparently directed against. By occupying the defunct Brent Spar oil rig and storage tank for three weeks in order to draw the attention of the world's media to Shell's unsustainable disposal practices, Greenpeace simply created a 'human interest story', a sense of high drama, in which the tension between the protagonists dominated the headlines. Consequently, issues about pollution and toxicity receded into the background, and nothing was done to question the rights of human beings to despoil the environment for the sake of greater profit margins. The mediatised recuperation of the Brent Spar protest, with its addiction to spectacle and theatre, accomplished no fundamental shift in ecological thinking, according to Kershaw. Instead, it merely continued the pervasive logic of 'business as usual': 'In this performance event, despite all appearances, the environment is a very minor player in a theatre ecology that seemingly cannot avoid being radically severed from the nature it so dearly wants to protect' (Kershaw, 2007: 266). Conversely, in Earth First!'s more oblique intervention, the attack on 'Smokey Bear' was read by spectators, Kershaw claims, as an attack on 'Nature' itself – something that, as he says, allows for more progressive modes of ecological perception in which the human moves out of the centre of the picture.

Theatre, Intimacy & Engagement: The Last Human Venue

While not perhaps immediately thought of as a work of ecocriticism – it is only belatedly that it has been referred to in this way – Alan Read's *Theatre, Intimacy & Engagement: The Last Human Venue* (2008) offers a profound and compelling theatre ecology.[41] Drawing on ideas from Jacques Rancière and Giorgio Agamben, amongst others, Read's work sets itself the important task of expanding what he calls the 'commons' of performance to include animals, children, and natural history (Read, 2020: 209–10). Counter-intuitively, the best stratagem for developing a new ecological 'ethics of association' (Read, 2008: 15), for Read, is to stop measuring what theatre can do and instead to highlight what it cannot, to stress its limitations:

> The tension between theatre and political claims has come about because most theatre is neither able to manage, nor to effect, the move from the specified to the specific, most theatre is in fact *singular* [...]. Theatre and the political are enemies because, despite everything Nicolas Bourriaud says in his much-quoted book *Relational Aesthetics*, the relationality expected of politics is betrayed repeatedly (different each night) by theatre's inability to relate. Theatre is dysfunctional, in this respect. (Read, 2008: 34; original italics)

The disfunction that Read insists on is central to the idea of theatre ecology that he develops in the final section of his book. Unlike the great majority of critics in Theatre Studies who advocate for an aesthetic that would bring spectators closer to 'Nature', Read is more concerned to acknowledge the essential 'unrelatability' of the medium. As he has it, there is something in theatre that troubles identity, an in-built rupture of sorts that calls into question the attempts of activists and ecologists to use theatre to establish proper connections between animals and people or 'Nature' and 'Culture'.[42] Theatre erases and suspends. It does not confirm essences or binaries:

> The relation between humans and other animals might unfortunately for the animal rights activists and ecologist, be less to do with the animals than the humans themselves, ourselves, the way that human agencies such as politics, ethics and law are already and everywhere suspended between man and animals. Here the experiment in question will always return to the renegotiation of what counts for human nature. (Read, 2008: 90)

For Read, theatre shows human beings as subjects 'without qualities', interminably caught up in a movement toward something, but never quite arriving there.[43] Counter-intuitively, theatre is not the 'last human venue' because it defines the human, but rather because it insists on opacity, a fundamental elusiveness, a blind spot in the human's being that can never be expunged or successfully treated. In its disclosure of non-coincidence, theatre transcends the actual and reveals, in its place, the indiscernible, presence of 'bare life' – life as an anonymous gratuity, the sheer biological fact of existence, its zero degree.⁴⁴ In order to unveil the neutrality of life, theatre must not make the mistake of creating empathy or identification with a character or even an animal. Rather, to reveal 'bare life' – what Giorgio Agamben, after Aristotle, terms *zoē* rather than *bios* – theatre has to deconstruct itself, to highlight its failure to represent the world positively.⁴⁵ Which is precisely why Read is so attracted to the work of Shunt, Goat Island, Forced Entertainment and Socìetas Raffaello Sanzio, all of whom insist on a subtractive aesthetics, a theatre that accentuates the necessarily mediated aspects of human existence, the obscurity of a fractured existence.

In a radical departure from Chaudhuri and Marranca and differently, too, from Kershaw, theatre's power, for Read, is discovered in its suspension of self-presence, in the fact that the living actor on stage is ironically undone by appearing as a body that seems so intensely close, so radically there.⁴⁶ Technological media are not so adept at disclosing this 'bare life', Read claims. Their use of flat surfaces 'screens out' the real by producing strong images – simulations – that allow the human to impose itself on the world as a non-alienated being, a subject with an identity. Once again, mimetic success is a tantamount to ecological failure:

> The theatre was the last human venue in as much as its objects were measured, not for their potential to act, but for their impotential to be realised. [...] Perhaps rather contrarily and disappointingly for the futurist technocrats, the mark of human nature at work in this venue (as distinct from professional success) was one of old-fashioned, pre-bionic *underachievement*, and it was this falling short, this remedial quality, not exemplary excesses that marked it out as peculiarly human, and peculiarly alienated. (Read, 2008: 4; original italics)

The remedial, pre-bionic status of theatre that Read alludes to above endows it with an uncanny quality – a hauntology – that pertains more to the future than to the past. In the theatre, the human subject is impoverished and weak, unable to surmount its alienation, devoid of all technocratic escape routes that would supposedly save it from its earthly fate to die out and end. Like those anticipatory spectres that haunt Thom van Dooren's writing on birds in *Flight Ways: Life and Loss at the Edge of Extinction* (2014) and Ray Brassier's post-solar worlds in *Nihil Unbound: Enlightenment and Extinction* (2007), Read's ghosts disclose the possibility of an extinction to come, the spectre of a world without us:

> The last human venue is the place of performance where such distinctions between humans and other animals are played out – a venue for the coming together of a repertory of self-evident facts where there is no knowing how the self will survive, and abundant evidence that it will not. (Read, 2008: 1)

The 'lastness' of theatre, its affiliation to a timeframe beyond the human, does not cause Read to assent to nihilism, however. Although extinction is surmised as a horizon that we may be unable to transcend, theatre's necessity is found, for Read, in its possibility for lengthening and deferring that end, for teaching us how to 'live well at last', as Jacques Derrida puts it. Such an impulse towards an ethics of hope resonates in Read's reference to the 'Lazarus effect' in the final pages of his book, a parable which emphasises the potential that theatre has to start again, to give birth to a second origin with life itself:

> The time that remains in this scenario would appear to be one of hope therefore. Having begun to measure these distances [to an end], having begun to act in the interests of their lengthening, the last human venue would appear to be a venue with a purpose, even if that purpose is still not political but finally at one with politics. If the word 'venue' in its original French form meant 'a coming', then this place would appear to be the assembly of those before any second coming, any messianic or metaphysical aspiration to transcend the materiality of a threat of an ending. (Read, 2008: 273)

It is a source both of regret – and of real bemusement to me – that Read's ideas have not found a wider audience amongst ecocritics in Theatre and Performance Studies. As well as linking politics to environment and expanding the 'performance commons', Read stresses that 'ecotheatre' ought to challenge the centrality, indeed the very ground – of the human animal – that figure whose power seems to go unnoticed in so much ecocritical commentary in Theatre Studies. Typically, the move is two-way. In Read's dogged refusal of synthesis and sublation, the 'lessness' of the theatrical medium becomes a 'more', its 'lastness' a way, finally, of humbling the human, of exposing it to the earth that it shares with other forms of organic and inorganic life, and which it can never fully transcend and/or manage to escape, despite the deranged claims of those ecomodernists who want to terraform the planet or exit it completely. By insisting on 'impotential', Read provides theatre with a disabused ecological sensibility – performance theory for the Anthropocene. This is an ecology in which theatre has the difficult task of embodying ambivalence, of teaching its audience to live without completion, of being permanently divided, out of step with itself. To be human, in Read's view, is not to know what one is; it is to exist as a stutter, an animal of a different kind that is never able to commune with 'Nature' or to abandon it, but is always caught up in its flow, its processes of becoming. Of the four thinkers I have discussed, Read is the most unsettling; the one whose ecological hope is provisional, contingent and always in doubt of ever arriving, precisely because the human itself is not just a 'paradoxical primate', as Kershaw maintains (Kershaw, 2007: 236), but a radically unknowable one, a *pharmakon*, non-coincident and vulnerable. In the face of Read's radical openness, no paradox or negentropy can save us; rather, we are left trembling on the banks of some extinction to come, at the mercy of a lastness, an indeterminate finality that needs to be embraced as the only way of lengthening our survival.

Reflections

As these parsings expose, probably for the first time, theatre ecology is not a unified field. It is multiple and contentious, constituted by divergent histories and opposed ways of operating. Putting it perhaps too schematically, Chaudhuri's method is to concentrate on interpreting and reading dramatic texts, even when her aim is to propose theatre as a phenomenological site for being together; Marranca's to attend to the formal possibilities of modernist and postmodernist scenographies and dramaturgies; Kershaw's to analyse the paradoxical affordances of performance events and site-specific interventions taking place outside the auditorium; and Read's to deconstruct theatre as a site for human appearing. Crucially, each of these methods reveals a very different concept of ecology. Chaudhuri's is focused on restoring post-Heideggerian modes of dwelling; Marranca's on biocentric holism; Kershaw's on a synthesis of cybernetics and phenomenology; and Read's on an ecodeconstructionist interrogation of the human, showing how the *anthropos*, like all creatures, is bound up with 'bare life', the thing or force that it can never coincide with nor separate itself from, no matter how hard it tries to do so.

Recalling my point about methods and praxis (see pp. 23–32), these different concepts of ecology are accompanied by equally singular reflections on what representation stands for and is supposedly able to do. Chaudhuri puts naturalism in question and argues for self-conscious and minimalist ways of staging ecology and environment; Marranca takes a more celebratory approach and reads theatre as a tool for 'cultivating' biocentrism (in the important double sense of cultivation); and Read is critical of any attempt to represent ecological issues directly, preferring a theatre that deconstructs and suspends the primacy of identity. Kershaw's position is arguably the most extreme – at least when it comes to valuing theatre as a site for showing. For Kershaw, as I explained on p. 43, the black-box stage is not a useful mechanism for 'doing' ecology, as Chaudhuri, Marranca and Read all implicitly affirm. Rather, as he sees it, the *theatron* itself is the problem, in respect to which – at least in its modern, industrial form – it teaches spectators, in the most insidious of ways, to constitute themselves as distant observers who place themselves 'outside' of the event they are actually 'in', like good Cartesian subjects. Given that his premise troubles the very existence of this critical field, it is odd that the logic driving Kershaw's provocative refusal of theatrical space has not been properly debated or contested.[47] Instead critics have either taken him at his word or else preferred to push forth regardless.

Similarly, no one, to my knowledge, has thought it important to contemplate the reasoning behind the decision of Marranca, Kershaw and Read to depart from the traditional styles of scholarly discourse. In each of their cases, critical engagement with the 'natural history of performance' necessitates recourse to a different register of expression, in such a way that academic distance, the quest for interpretative objectivity, is disrupted and forced to change. Marranca's style is evocative and lyrical; Kershaw's conceptually inventive and autobiographically attuned to the role of place in practice-based research; and Read's is playful and ruminative, mixing humorous digression with high theory. When it comes to theatre ecology, there seems to be a desire, a need even, to get beyond interpretation, to show evidence of a more fundamental displacement in how the critic relates to their theme, situates themselves with respect to performative things, material and events.

If one is committed to the ecological work that theatre and performance can do, the methodological and stylistic questions that Chaudhuri, Marranca, Read and Kershaw have left us with need to be examined, their epistemological, ontological and aesthetic assumptions interrogated and unpacked. Not to do so is to run the risk of irrelevance, to continue with practices that may possibly be out of line with the present state of the world or undermined by hidden contradictions. As Bergthaller cautioned in the opening pages of this chapter, faced with the uncertainties of climate change we need new concepts of 'Nature' – ones that are suitable for the task at hand. The appeal is made to all theatre scholars and practitioners. As I have already said, the ideas of Chaudhuri, Marranca, Kershaw and Read are not confined to their own texts alone. Their interventions have served as essential reference points and primary foundations for a nascent sub-discipline within Theatre and Performance Studies. To interrogate their methods is, then, to interrogate how the idea of ecology is theorised, practised and imagined by performance scholars and practitioners today – an investigation that allows for a disciplinary inventory to be made, a critical genealogy sensitive to affordances and aporias and attuned to questions of inheritance and legacy.

Alan Read gets to the crux of why such an expansive or generic interrogation is required, when he makes the discomfiting provocation that 'there is no more point in throwing ecology at theatre and hoping for the best, than there is in prematurely binding theatre and the political or social issues and performance without addressing the serious problems of their more substantial relations' (Read, 2008: 250). At a time when work in this area is starting to accumulate and increase, it seems propitious, if not vital, to reflect on Read's challenge and to ponder just what a 'more substantial relation' between theatre and ecology may consist of. The trap to avoid, as a matter of urgency, is the assumption that there already exists some blueprint or model for doing theatre ecology, which can be uncritically rolled out and applied. Everything is to be invented, again and again. It ought to come as no surprise that the genealogy I am tracking is inevitably partial. There will be gaps and others may take issue with what I have selected and omitted. That is exactly as it should be. The idea for a theatre ecology that I am proposing in this book is open to revisions, a mere starting point. It knows it cannot say everything. The imperative is to produce debate, to think critically and creatively about theatre's possibilities for constructive ecological interventions to come.

Hermeneutics: textual ecologies

In light of its militant call to read *for* 'Nature' in dramatic texts, it is no surprise to find that Chaudhuri's hermeneutic approach to ecology has proved the dominant methodology in Theatre and Performance Studies to date – the one that seems most immediately self-evident and tangible. Derivations of her interpretative method have resulted in important revisionist accounts of the western canon, such as Downing Cless's attempt to unearth an occluded environmental history in theatre from Aristophanes to Brecht in *Ecology and Environment in Western Drama* (Cless, 2010); Carl Lavery's and Clare Finburgh-Delijani's more delimited excavation of a nascent ecology in modern western theatre in their collection *Rethinking the Theatre of the Absurd: Ecology, Environment and the Greening of the Modern Stage* (Lavery and Finburgh-Delijani, 2015); and Theresa J. May's revisionist history of the environment in modern US drama in the monograph *Earth Matters on Stage* (May, 2020). Whereas these publications read classic texts against their ecological grain, studies such as Marissia Fragkou's *Ecologies of Precarity in the Twenty-First Century Theatre: Politics, Affect, Responsibility* (Fragkou, 2019); Vicky Angelaki's short book *Theatre and Environment* (Angelaki, 2019); and Mohebat Ahmadi's PhD *Towards an Ecocritical Theatre: Staging the Anthropo(s)cene* (Ahmadi, 2017) follow a more frontal line of investigation.[48] Fragkou reads a number of modern British playwrights through a lens of precarity; Angelaki provides an excellent and succinct overview of ecological theatre in the UK, US and Europe with an emphasis on 'accounting for the possibilities and modes of positive present intervention' (Angelaki, 2019: 75); and Ahmadi looks at the responses that established playwrights in the West (Andrew Bovell, Caryl Churchill and Chantal Bilodeau etc.) have made to the ecological crisis in the Anthropocene.

There are also, as one might expect in this critical tradition, numerous stand-alone contributions, often found in edited collections and special editions of journals. Prime examples include Elaine Aston's (2006) early ecocritical analyses of the plays of Caryl Churchill – a writer whose play *Far and Away* (2000) is already an ecodramaturgical 'classic';[49] Theresa J. May's analysis of Marie Clements's *Burning Vision* (May, 2010), the essay that introduced the influential notion of ecodramaturgy to theatre criticism; Stephen J. Bottoms's 'Climate Change Science on the London Stage' (Bottoms, 2012), a critical survey of the sometimes problematic attempts of contemporary UK playwrights to create dramatically coherent 'cli-fi dramas'; Graeme MacDonald's generative introduction to the 2015 Bloomsbury edition of John McGrath's *The Cheviot, the Stag and the Black Black Oil*, which brings Theatre Studies into alignment with the Energy Humanities due to its focus on infrastructure, oil aesthetics and 'extractionism'; Una Chaudhuri's analysis of Wallace Shawn's play *Grasses of a Thousand Colours* (2009), which argued for new forms of 'incommensurability' – a 'drama of bad ideas' – to reflect the derangements of the Anthropocene (2016a); and Trish McTighe's queer ecological take on the presence of non-human objects and vibrant matter in Doug Wright's play *I Am My Own Wife* (McTighe, 2017). While many of the texts complicate naive concepts of ecology that would simply equate it with representations of 'Nature' alone, curiously none of them with the exception of Cless, who discusses his own ecodramaturgical stagings of classic plays, stresses the fact that reading a dramatic text is very different from reading a literary one, something with its own materialist and affective purchase.

Because of that absence, there has been no real engagement with theatre as a corporeal medium in this text-focused methodology. As a consequence, the exciting possibility that Chaudhuri held out for thinking of the stage as a new collective *oikos*, a phenomenological site for communal listening in her response to Spalding Gray, has been frustratingly obscured. Indeed, in most of the critical texts mentioned above, the theatre is rarely, if ever, considered as a materialist space in its own right,[50] a milieu where actors are part of an ecological event which affords or constricts their agency and subjectivity; and neither is there any sense of the critic's body being affected by the images or words impressed upon it, made 'viscous', to transpose a sensate, queered word from Nancy Tuana (2008: 189).[51] In this critical model, ecological exegesis – not praxis – is the order of the day.[52]

Extrapolating from the contradiction that Kershaw had spotted in the epistemologies of Chaudhuri and Marranca, hermeneutic readings of performance ironically tend to affirm a kind of placelessness – the 'view from nowhere' – that their arguments are expressly constructed to counteract. At the same time, they also perpetuate a logocentric and/or correlationist view of humanist subjectivity, in which the subject is not only divorced from their environment, but untroubled by the acts taking place in front of or around them, secure, as they seem to be, in the power of language and interpretation to remain intact, and, presumably, to lead us out of the epistemological and ontological disarray we are in.

Unless one thinks, *contra* Marx, that it is enough to interpret the world, then it is difficult to see how this textual method for 'doing' theatre ecology can affect a radical shift in how spectators might relate differently to the earth. More perversely still, and as I have been intimating throughout this section, reading for ecology in dramatic texts results in a decidedly anti-ecological mode of perception, in the respect to which the affective charge of performance, its capacity to appeal to something beyond the *anthropos*, is kept at a distance in order for the critic to enact a quasi-surgical operation on it. This is not criticism that brings us down to earth. In actuality, it is designed to protect us against corporeal sensations, thus perpetuating what Peter Sloterdijk, in his environmentally inspired critique of spherical thinking, designates as 'immune-systemically effective space creations' (Sloterdijk, 2011: 28). So while there is no doubt that the ecohermeneutic methods used by critics who have followed in Chaudhuri's wake have certainly expanded the prominence of a certain idea of ecocriticism in Theatre and Performance Studies, it is a scholarly practice that is beset with contradictions and dangers that one would do well to recognise. As I have suggested, it is not just naturalism that perpetuates idealism under the cover of empiricism, as Chaudhuri so cogently points out. Ecocriticism that adopts a literalist approach has a similar tendency.

Site-specifics

On account of the contradictions inherent in the textual approaches listed above, a number of scholars and practitioners have been keen to experiment with Kershaw's alternative paradigm of theatre ecology. Significantly, though, apart from Judie Hudson who has very usefully transposed Kershaw's ideas on systems theory to the performance event itself (see pp. 29–30 and p. 69, note 10 in this book), the 'Kershaw' in circulation in Theatre Studies has not been the theorist of paradox, weakness and negentropy – the one who brilliantly undermines anthropocentric ideas of intentionality and volition. Rather his influence has been most keenly felt in the preponderance of environmental work rooted in corporeal engagements with actual sites, landscapes and places. This concern to place bodies in direct contact with 'Nature' or to immerse them in 'vibrant matter' (Bennett, 2010) has been loosely characterised by three different tendencies in subsequent scholarship.

First, a desire to produce more embodied modes of human perception, in which scholars draw on the discourses of ecophenomenology, new materialism and object-oriented ontology in order to show how site-based performances can entangle spectators in the 'more-than-human world'. Second, a commitment to making activist interventions in real places, be that in toxic sites, ruined buildings, agricultural businesses, contested spaces, city streets, museums and galleries.[53] And third an attempt to engage with the environment through practice. This research methodology has proved increasingly popular in the past decade or so, particularly in Anglophone and Nordic countries, and numerous artist-scholars have drawn on a diverse range of ecological discourses as a way of affirming the specific environmental work that theatre and performance practices can purportedly achieve.[54]

Notwithstanding the valuable insights that these different forms of site-based research have produced for ecocriticism, there are nevertheless specific problems with each of them that need to be addressed in their own terms. The rationale, for instance, that some scholars and practitioners have given for exiting the 'all-too-human' playhouse has often been predicated on a belief in performance's capacity to somehow afford an unmediated encounter with 'Nature', 'matter', 'landscape', 'place'. The difficulty with this position is its assumption that the real is found outside, in the space beyond the artificially, air-conditioned walls of the black box. In investing in such a logic, there is a danger that a new binary opposition might reassert itself between 'Nature' and 'Culture'. Here site runs the risk of becoming the unmediated home of what is thought to be proper and true, an idea or *logos* that replays the exclusions of modernity in so far as it posits origins, hierarchies and oneness as absolute values.

While I certainly endorse the search for intense, sensate modes of relating to the earth, my worry here is that to encourage, as Kershaw does, immediate contact with the outside is to forget that theatre is always already, in its very operations, a non-human medium, a configuration of 'spacetimematter' in which bodies are opened up to terrestrial and cosmic forces that know nothing of insides or outsides. Despite the initial oddness of this proposition, perhaps, this is an environmental viewpoint that a more materially minded generation of critics in Theatre and Performance Studies such as Joslin McKinney (2015), Martin Welton (2018), Eddie Patterson

(2019) and Sarah Lucie (2020) all share. For these critics, as for me, theatre is already an environmental site, a place to be weathered, an atmosphere to breathe in. Conceived as a cosmic house, a 'techno-nature assemblage', theatre is not just a locus for disembodied looking. On the contrary, it is one where the body is conceived as agent of connection, the eye an organ for sticking to things, for being enervated. This alternative approach to looking reveals a blindspot in Kershaw's thinking. The thing that Kershaw rejects about the theatre space – the idea that it leaves one separated and distanced – is predicated on the very principle that his attachment to Bateson's second-order theory of cybernetics ought to insist upon: namely, the fact that no observer can stand apart from the system they observe.

An additional but different dilemma that besets the quest for ecological experience in much site-based work concerns the lack of attention given to political economy. Although there is certainly an explicit and often intersectional concern with issues of disability, gender and queerness in the practice and scholarship of site-based ecocritics, there has often been a neglect of historical materialism and political economy, as Rebecca Schneider pointed out in her short but perspicacious critique of how new materialism has been deployed in Theatre and Performance Studies (2015).[55] Indeed, some scholars interested in place and landscape have tended to ignore socio-economic factors altogether, preferring to engage in what Jane Bennett has termed an ethics of enchantment, the inspiration for a type of 'wonder' that produces 'actual attachments' to one's surroundings, be they rural or urban, without ever saying how such experiences enable the move from individual to collective action or from the local to the planetary (Bennett, 2001: 162).

The drawback with this ethical understanding of re-enchantment is not only that it bolsters the creative power of the human subject in ways that often underestimate the agency of matter itself – the capacity of the earth to turn against the *anthropos* – it also tends to forget the very real plight of the planet today: the sense in which the soil or ground is being blown away or made toxic by industrialised agricultural practices.[56] However, on a positive note, it must be said that an increasing number of ecocritics in Theatre and Performance Studies, influenced by decolonial thinkers such as Marisol De La Cadena (2015), Zoe Todd (2016) and Déborah Danowski and Eduardo Viveiros de Castro (2017), are beginning to reconfigure the idea of site to include such things as economics, land management, extractivism and agrilogistics. In these doubly materialised and political versions of site-specificity, place is no longer about the solitary production of sensuous immersion; rather, it is to do with the conjuring of repressed imperialist memories, the terrible devastations, exhaustions and violence wreaked by settler communities on Indigenous bodies, minds, soils and animals. For theatre scholars such as Helen Gilbert (2013), Lara Stevens (2020),

Chris Bell (2020), Gabriel Levine (2020) and Michelle Nicholson-Sanz (2020), 'Nature' is always shot through with postcolonial and decolonial histories and politics, and thus open to contestations and dissensus.

But even here, in site-based scholarship that is so rigorously attuned to the toxicity and injustices of capitalist imperialism, difficulties still remain. Only now the issue is to do no longer with the supposed absence of the human, but with its excessive presence, the tendency to focus on human agency to the detriment of the planet's agency. Too often in its admirable attempt to do something, site-based eco-activism remains tied to a restricted ecology, one that is undone by its humanist faith in intention and volition, as Kershaw himself was quick to realise. In the impatience of activist practitioners and scholars to do something, to make an intervention, the ecological values that Kershaw draws such astute critical attention to – irony, paradox, and obliquity – are conveniently dispersed with, and the aesthetic becomes an instrument that can be brandished in an economy of ends and means. Once again, the *epistēmē* that produced ecological devastation and ecosystem collapse in the first place evades critique. Instrumentalist reason remains central, sometimes violently so. By consciously seeking change, in demanding results, eco-activism overlooks what I called the alternative activism of the artwork in my comments in the *Polemos* chapter of this book (see p. xii): the paradoxical sense in which theatre has the ability to produce new, non-instrumentalised ways of being on the earth that may not – at least in the immediate context – *seem* to serve any purpose. If we are to have what the Martinican scholar Malcom Ferdinand terms a 'decolonial ecology' in Theatre Studies (Ferdinand, 2022), then the way to do that is to bring aesthetics, ecology and politics together, to remain committed to the borderline that would allow them to exist in unison, if not in unity. However, the 'fracture' between environmentalism and colonialism that Ferdinand's decolonial ecology seeks to overcome must not be too neatly sutured, the crack must remain visible (Ferdinand, 2022: 3).[57]

My critique of Kershaw's final category, practice-based research, is of a different but related order. Here I take issue not with the methodology in itself – I am completely convinced that ecologically informed practice has the capacity to produce new knowledge of what it means to be a human on the earth – but with the epistemological logic that often characterises the reflective component of the research. Referring back to my earlier point about stylistics and metaphysics (see pp. 17–18), I am often surprised by the fact that ecological work which claims, so assiduously, to trouble the centrality of the subject should mobilise a form of language that is so masterful and neat, so in total control of its processes. Just as curious, in this respect, is the attachment that so many posthumanist practice-based researchers have to forms of autobiographical writing, in which the researcher, even when admitting epistemological defeat and ontological failure, has such easy recourse to the personal pronoun, the 'I' that surveys and appropriates all that it sees and touches. For, as I will explain in Chapters 2 and 6, to be affected by matter is precisely not to encounter feelings and emotions that are already known by a subject confident of its 'ground'; it is to be capsized by excessive affects and impulses that have no name and that language struggles to express and designate, principally because they happen all at once, at a rhythm and pace that neither perception nor words can keep up with.

While I am all too aware that one is always compromised by the limits of language as well as by the frame of one's humanness – there is no direct access to things – there are nevertheless ways of writing that can *express* what human beings are unable to articulate semantically, that can evoke the disturbances of matter, the impress of non-verbal affects. What I am arguing for, here, is a language of ellipsis and reticence, a tone of voice or syntax that would place the 'I' in question, undoing its mastery, transmitting the excessive life force that inhabits all bodies. Wallace Heim does this beautifully and elegantly in her article 'Can a Place Learn' (2012b) by simply asking questions, by admitting unknowing, by circling around her own hesitation. In similar vein, Lisa Robertson argues for a practice of critical writing founded on the 'failure of transparency', an 'inaccessibility [...] where perception disperses identity in a movement toward unknowing' (Robertson, 2012: 16).

In this way, through this writing, ecological practice is suffused with the earth as a non-human force, as something that is very much there but which refutes direct representation, simple correspondence with a conventional signifier. When this struggle to express the 'outside of thought' in practice-based commentary is missing, and replaced with explanation, description and rhetorical exposition, the ecology of the practice itself becomes suspect, the mere illustration and/or application of an idea imported from without, and not, as it is often advertised as being, a discovering through doing, an effect of immanent experimentation. If the aim is to demonstrate how practice can provoke novel forms of ecological thinking and feeling, then surely it is not enough to keep extant ideas and vocabularies in place. Rather, ecologically orientated practice-based research ought to interrogate the validity of all theory extraneous to the artwork, to make conceptual thought acknowledge its limits, to wager on the terrestrial quality and value of the artwork. Unfortunately, much site-based practice that draws on ecocritical theory does not do that, and so misses the opportunity for forging genuinely new insights and languages that might allow spectators to get a new 'taste' for the earth through performance.

In making these criticisms of the methods advocated by Chaudhuri and Kershaw as well as the works of critics influenced by them, I am not claiming that every piece of research that deals with site or reads dramatic texts is subject to the same problems and contradictions alighted upon here. Rather, the point is to gesture to generic gaps in these methods that need to be acknowledged and worked through. From my perspective, the main problem with the frontal approaches to ecology and environment that I have mentioned is that despite their apparent differences – one is focused on the text; the other on diverse approaches to site-based performance – they both tend to uphold a conventional distinction between self and other that underpins the disastrous anthropocentrism so entrenched in humanist thinking. In both approaches, the human subject is endowed with a preformed, substantive identity that supposedly separates it from the earth, even when it looks to return to it, as both Chaudhuri and Kershaw do, albeit in contrasting ways. The desire to encounter what Kershaw refers to as 'nature's "nature"' no longer makes sense today (Kershaw, 2007: 319). Not just because the human is always already 'Nature', as both Chaudhuri and Kershaw have argued in their respective writings on animals, but because 'Nature' is always already 'Culture', informed by codes, writing systems and processes that divide it from itself. As I will discuss in more detail in Chapter 2, this unsettling, posthumanist view of 'Nature' entails new methods of working, approaches that trouble the epistemological and ontological assumptions of dominant ways of doing ecocriticism in Theatre and Performance Studies. This is because posthumanist ecology posits the Nature/Culture relationship in terms that neither Chaudhuri nor Kershaw does. That is to say, it is figured as a disjunctive unity, a constantly moving borderline that is always 'more and less than one', an excess or abundance that refuses to be known and/or experienced by a phenomenological subject set on reestablishing some 'proper' relationship with it. In posthumanist ecology, neither propriety nor proximity is a viable option. There is nothing to repair or restore, only ways of constructing provisional relations, of living undone in 'impossible intimacy'.

From this point of view, it is regrettable that Kershaw's championship of 'paradoxology' has not been reflected on in greater depth or put to a different use (Kershaw, 2007: 98–132). For there was – and still is, I believe – a real possibility, here, to advance a sophisticated but counterintuitive type of theatre ecology that would think the earth in such a way that it cannot be mastered or put into signs. To cite Cary Wolfe's reading of Niklas Luhmann and Wallace Stevens in the essay 'The Idea of Observation at Key West: Systems Theory, Poetry and Form beyond Formalism' (Wolfe, 2010), paradox makes the non-human world (un)available in such a way that it refuses to submit to the intentions and fantasies of the human subject. In doing so, paradox allows the earth to *exist as earth*, something that is *'incapable of emendation'* (Luhmann in Wolfe, 2010: 276; Wolfe's italics).

In this paradoxical pedagogy, neither artist nor critic is concerned with the dissemination of ecological meaning. On the contrary, they both look to gamble on the intrinsic value of aesthetic experience, in allowing the work to be like 'Nature', sufficient to itself. In opposition both to activist artists and to funding councils that are desperate to be seen to make us care, the discomfiting irony, here, is that the more we stop trying to care for the environment and pay attention to the artwork's reticence to show or tell, the more ecologically conscious we become, better able to escape the tragic paradox that so often besets ecologically engaged work: namely, that its concern to mobilise bodies merely repeats the militaristic logic that is only ever concerned with fulfilling its own agonistic ends, quite literally failing to see the 'wood for the trees'.

Natural histories

In comparison to the ubiquitous presence of both textual and site-based approaches to ecology in Theatre and Performance Studies, the more indirect, non-representational methods advanced by Marranca and Read have played a relatively minor role. Until recently only Stephen J. Bottoms's and Matthew Goulish's experimental collection *Small Acts of Repair: Performance, Ecology, and Goat Island* (2007), with its collage-like dramaturgy, came anywhere near to tapping into the type of 'natural history of performance' advanced by Marranca and Read. But there are real signs that this neglect, thankfully, is beginning to be addressed, as scholars and practitioners such as João Florêncio (2015), Augusto Corrieri (2017), Sylvia Battista (2018), Christel Stalpaert (2018), Kristof Van Baarle (2018), Maaike Bleeker (2020), Hannah Kaya (2020) and Manuela Infante (2020) are starting to rethink what a simultaneously more materialist and aesthetically abstract relationship between theatre and ecology could be.[58] Such non-representational work promises to reconfigure the entire field ontologically and epistemologically. And one wonders, as I have already hinted at on pp. 53–6, if, along with a decolonialist turn, this is where one of the futures of theatre ecology might reside. In the attempt, that is, to understand how the appearance of human and non-human bodies in a shared time and space allows for the production of ecological becomings that do not need to conform to themes or stories, and which concentrate on the immanence of the theatre event itself. There is a fundamental queering at work in this approach, since these scholars realise that terms like materialism and abstraction, 'Nature' and 'Culture', and form and content are always already artificial, liable to shift and drift, to find themselves imbricated in the place of the Other they *only* seem to oppose.

To make the 'queer' claim, as Marranca and Read do, that theatre history is always already a site of natural history means that there is no need for theatre to represent the natural world or to abandon the auditorium. When theatre renounces Aristotle's plot-driven *mythos*, and instead embraces the spatio-temporal conditions of its own *opsis*, a new set of ecologically determined questions start to haunt – and thus internally displace – theatre's inbuilt humanism. The agonistic borderline that Chaudhuri and Kershaw, despite their apparent antagonisms, were desperate to overcome by positing 'Nature' as something to return to and find union with, is no longer an issue from this point of view. In a series of provocative questions Marranca, for instance, proposes a very different understanding of naturalism from Chaudhuri's, one in which 'Nature' and 'Art' form part of a common biosemiotic process, a single monadic substance in which everything interacts and impresses on everything else:

> How do geography and climate influence a work? What are the ways in which plant and human life, animate and inanimate entities, the natural and artificial interact? How do biology and the body determine the human drama? Frequent travels abroad influenced this line of enquiry as I followed the expression of culture in different kinds of landscapes, the quality of air and light and water and wind: the design of doors, windows, parks, and textures of stone and wood. What effect do they have on the performing body, the perception of objects in space? And what of the diverse species of birds and plants and the spectacle of contemporary life in the new ecologies of cities? (Marranca, 1996: xiv)

For Marranca, theatre ecology is not about healing severances. As her questions demonstrate, 'Nature' is already implicated in how humans make and perceive their worlds, identities and aesthetic exchanges. To fuse Morton with Bateson, Marranca adheres to what could be termed 'a queer theatre ecology of mind', a naturalised ontology, characterised by 'a non-hierarchical embrace of the multiplicity of species and language in a work' (Marranca, 1996: xvi). Differently from the frontal model of ecodramaturgy advocated by Wendy Arons and Theresa J. May, Marranca's theatre ecology refuses to place the 'environment' centre stage as an object to represent or outside to index. Rather, Marranca offers Theatre and Performance Studies a different kind of ecomimesis, an expression of the earthly processes and systems that any performance is already entangled with. The aim is not to provide knowledge, but to transform theatre into a 'bloom space', a garden for spectators to visit and enjoy (Marranca, 1996: 21). Like John Cage, Marranca treats her audience 'more as *ecological* than *sociological* fact', participants to 'grow' rather than pupils to teach and inform (Marranca, 1996: 30; original italics).

The ontology of Marranca's theatre garden is multiple and expansive, a field composed of humans, animals, plants, machines, technology. The environmental legacy of Marranca's methodology today is best seen by moving away from nature-based models of ecocriticism, and instead engaging with the generative but latent ecologies inherent in the work of scholars such as Andrew Sofer (2003, 2013), Nicholas Ridout (2006), Jennifer Parker-Starbuck (2011), Peta Tait (2012), Laura Cull (2013), Marlis Schweizter and Joanne Zerdy (2014), Mischa Twitchin (2016), Peter Eckersall, Helena Grehan and Ed Scheer (2017) and Nicolás Salzar Sutil (2018). These theatre thinkers challenge the primacy of the human subject by concentrating, in various ways, on its creaturely life, prosthetic dependencies and cyborgian intersections. In their publications, theatre is an aesthetic space where exceptionalism is questioned and the 'human animal' posited as always already *in medias res*. Whether they are explicit about their environmentalism or not, scholars interested in robots, cyborgs and information technologies are all engaged in an expansive practice of ecological thought – a way of relating to and perceiving the world as a series of 'more-than-human' encounters where becoming is catalysed through the presence of objects, things and systems.[59]

From these posthumanist perspectives, there is no need to flee 'darkened auditoria for a more authentic form of participation in 'Nature' that site-specific or participatory theatre supposedly provides. A more apposite way of operating is to explore how performance already stages the earth, via the material and immaterial relations that it sets up, confounding, therein, any hard and fast distinctions between theatre and 'Nature'.[60] This theatre ecology is an 'ecology without nature', as Morton puts it, an ontology and epistemology of intra-actions, something reticular and entangled (Morton, 2007).[61]

Read intensifies these posthuman entanglements by contesting what is arguably the most sacrosanct of all boundaries: the one separating life and death.[62] According to Read, as I explained on pp. 47–50, we do not go to the theatre to see what the human animal is (*bios*) but to come into contact with the anonymous 'bare life' (*zoē*) that humans share with all forms of organic and inorganic matter. Where western humanism looks to define what a proper human is, theatre reneges on this ontological and epistemological desire. In a sense, it is an 'idiot' medium, a technology of dis-appointment:

> Neither the human life nor the animal life is ever going to appear there, but only 'a life' that is separated and excluded from itself. It is 'bare life' that has finally appeared after all this machinery's grinding and whirling, and it is performance that nightly in the human laboratory, that is, in the last human venue, has demonstrated the working of this device. (Read, 2008: 96)

Read's dismantling of what he calls, after Giorgio Agamben, theatre's 'anthropological machinery' (Read, 2008: 8) marks a key moment in theatre's ecology that is currently under investigation by critics such as Lourdes Orozco (2018), Felipe Cervera (2019), Kristof Van Baarle (2020) and Sarah Lucie (2020), all of whom are interested in extinction events, those strange endings that no Apocalypse can redeem through the consolatory imposition of a denouement. For what is ultimately being evinced by Read's focus on *zoē* is a new way of thinking through the relationship between 'Theatre' and 'Nature' that fundamentally disturbs all extant theories of theatre and ecology, even those like Marranca's that refuse to sever 'Nature' from 'Culture'. Read's last human venue does not ask spectators to overcome their isolation by entangling themselves with external substances and landscapes beyond the self; rather it shows ecology to be internal to the human, a pre-individual wound that cannot be healed, an anonymous force that fractures and alienates.

As Read's analysis of the attempted resurrection of Roger 'Rogue' Riderhood in Charles Dickens's late novel *Our Mutual Friend* (1865) demonstrates (Read, 2008: 86–7), bare life exceeds the *bios* of the human; it carries on, individuating further, finding new bodies, organisms and viruses to host it, to give it temporary shelter.[63] Where Marranca's environmentalism is predicated on oneness and communion – a holism that hampers its ecological potential as I point out in Chapter 2 – Read concentrates on elusiveness and division, an ontology characterised by, on the one hand, the impossibility of ever being whole, and, on the other hand, the very real possibility of human extinction.

Read's theatre ecology is disquieting; it asks difficult questions. And, because of that, it is vital that more attention should be paid to it. In its dislocation of the subject, its concern with the invisible, absent presence of bare life, Read highlights the sense in which the human subject is undone by the very thing that constitutes it: life itself. The last human venue does not offer consolation. Yet neither does it condone some leap into nihilism. Instead, it attaches itself to the present participle, the tense of the continual present. The present participle expresses an ambiguous temporality where an endpoint is simultaneously moved towards and yet also deferred. In that deferral, that openness, something else is permitted to emerge, a new adventure in finitude, different ways of inhabiting the planet. Which is perhaps why Read, for all his lucidity about endings, sees theatre as performing the Lazarus effect, the miraculous 'place where you are given the chance to begin again' (Read, 2008: 279). Crucially, though, and here I insist on a meaning that Read's Biblical reference appears to leave ambiguous, Lazarus's rebirth has to be seen as terrestrial. His return must tie humans to the earth, not encourage them to exit it through some act of spiritual transcendence. No God can save us, not even a Messianic one.

If it is to make good on its ecological potential, Read's methodology needs to be better 'grounded', in all senses of the word. Despite his generous references to contemporary practitioners, there is no analysis – at least no close, granular analysis – of how bare life is composed in their work, and thereby no attempt to express how spectators are affected experientially by coming into contact with it as living bodies. Precisely because, as Dominic Johnson says in a perceptive review (Johnson, 2008: 514), Read's thinking and writing are so intellectually rich that they run the risk of overlooking what one could call the haecceity or thisness of theatre – that thrust of matter that exposes us to the invisible affect of bare life (see pp. 65–7 in this book). There is, then, a *décalage* or gap between what is being argued for discursively and what is being transmitted in and through Read's analysis. For all his brilliant and often hilarious stylistics – jokes, 'allusions', lines of inspirational flight – Read ultimately clings too tightly to the concept, to intelligence, and so the intensity caused by the 'word-blind' flux of the theatre is tempered in his account. Integral as it most certainly is – as it has to be – to any practice of theatre ecology today, what we are left with in Read's writing of ecology is the 'after-image' of an event, the distillation of his thought, not the shock to it.

A possible way of responding to that absence of sensation, to the stuttering that theatre imposes, is to return to the moment of the performance event itself, to describe, carefully and with precision, its modes of composition, as Marranca does in her detailed portraits of Robert Wilson, Heiner Müller and Rachel Rosenthal etc. Yet in proposing that more granular move, vigilance is still required. Although Marranca is concerned with specific performances, with real stuff, she, too, cannot help but set herself up as a distanced spectator, someone who digests thought rather than catching it in 'the in-act', as Erin Manning encourages us to (Manning, 2016: 25). In Marranca's lyrical mode of ecoperformance analysis, the eye, as Kershaw was half-right to say, remains an organ that thinks, a disembodied consciousness. The difficulty with this practice of ecospectating is that the concept triumphs yet again, and so Marranca's readers could be forgiven for thinking that ecology is equivalent to the recognition of a hidden geometry or visual pattern in the work, not an affective force that is expressed through it and which troubles the subjectivity – and thus the intellectual power – of the critic or spectator to recognise or synthesise anything.

As I will argue in Chapters 5 and 6, to get closer to the troubling power of non-human sensation – to experience and write it – requires modalities of spectating and performance analysis attuned to the theatrical medium as a materialist apparatus whose *break through* is predicated on a *breaking down* of the discrete space that separates stage and auditorium, and in which the inarticulate body of the spectator is foregrounded, its confusion expressed in signs. What I am suggesting then – and this is what brings me close, ultimately, to the affective ecomaterialism of scholars such as Battista, Lucie, Patterson and Welton – is that theatre ecology ought to situate itself not just *within a* milieu but *on a* 'plateau', a word that in French combines theatrical and geological meaning, since it is both a stage and a continental 'shelf' or 'plate' (for more on this see pp. 189–94 in Chapter 6 of this book).

In the theatre ecology I am concerned with in this book, theatre is a dynamic site where bodies participate in a vibrational, rhythmic process that disturbs rigid distinctions between subject and object. To be drawn into this space is not to contemplate the biocentrism of the stage lyrically, as Marranca does in her view on landscape theatre, and neither is it to read dramatic texts with Chaudhuri's negative ecological history of modern drama in mind. Equally, it has little in common with Kershaw's and Read's strategies to apply ecological and philosophical ideas to the stage from the 'outside in', to wager on hope, no matter how tempered. Rather, I want to think about theatre ecology in more processual and im-mediate terms as a material practice whose ecological affordances stem from the type of 'inter-ested' thinking and feeling generated by the organisation, distribution and 'clash' of theatre things in some always impossible here and now.[64] Redemption is not the primary goal, here. Acceptance and attachment have taken its place, and there is an affirmation of a limited sense of human agency – one that is willing to let go of a consolatory transcendence that would somehow redeem the subject, recover a lost plenitude or point to a solution.

In the 'plate tectonics' of the stage, the eye is figured as an agent of touch, disturbed by and responsive to the bare life that circulates through, in and around it. The eye, here, is more like a lung than anything else. It conjoins *and* severs, contaminated by the atmosphere it always already 'breathes' in.[65] As Deleuze and Guattari remind us:

> We are not in the world, we become with the world; we become by contemplating it. Everything is vision, becoming. We become universes. Becoming animal, plant, molecular becoming, becoming zero. (Deleuze and Guattari, 1994: 169)

Against simplistic immersive and/or phenomenological accounts of participation, this molecular vision, as Deleuze and Guattari have it, is proximal, never immediate or one, a becoming with, not a dissolving into. It depends upon a minimal sense of distance being maintained, in a way that is close to but ultimately different from the type of non-relationality that Read insists upon. While I endorse Read's point that to want to participate closely with things, to seek to bring them close, is to engage in an act of appropriation or capture, in which the object loses its alterity, swamped, as it is, in a humanist logic of the same, I am dubious about how Read's take on distance divorces the human from the earth, making it too discontinuous, too lonely.[66] By contrast, to insist, as I do, on 'minimal distance' is to argue for a discontinuous continuity, to reposition the human within a plane of existence, a flow of bare life, that it is never entirely separate from nor fused with.

In 'near-distance' or 'far-closeness', the human is implicated in a terrestrial mode of relationality where bodies and objects radiate with an im/material haze, ecstatically streaming beyond themselves and yet retaining their singularity in that heterogenous flux. Anticipating what I argue for in Chapter 3 in a section on Guattari (see pp. 113–16), to exist in this indistinct and murky atmosphere, is to find oneself caught up in a process of eco-aesthetic transference whereby subjectivity bifurcates from itself and takes off on what Jeffrey Jerome Cohen and Lowell Duckert call 'a veer ecology' (2017), an ecology where one is connected and disconnected at the same time, always in the middle of things, on some transversal trajectory.[67]

To grasp the ecological potential of this veering, conventional ways of thinking about theatre and ecology need to be abandoned, along with rote concepts such as 'ecotheatre', 'ecodramaturgy', 'ecoscenography' and 'Anthropos(c)enes'. The aim is not to carve out a delimited niche, or to imprison theatre ecology within a genre or neologism that one can fix, appropriate and market. Rather, the imperative is to ecologise theatre, to fashion a *style* of analysis that is, at once, acutely attuned to pragmatics, ready to account for the ecological work that performance can do, while, at the same time, remaining cognisant of the fact that there is always an excess or surplus to theatrical experience that can never be articulated. In my mode of ecological analysis, something always escapes the spectator, a non-human force that disrupts its knowledge claims, that eludes its critical grasp, that messes with its habitual methods and methodologies.[68]

To show fidelity to that surplus, theatre ecology demands a style of writing that is infected by what it is unable to speak, that bypasses the 'I' in the very act of using the 'I', that is energised, in other words. To tap this impossibility, this abundance, is to give theatre ecology a bodily language, to show by doing, to admit failure. That I am unable to resolve these tensions in good scholarly fashion is an indication that I am on the right track, or so it seems to me. To affirm the earth is not to know what it is. Yet this wager on what Stefano Harney and Fred Moten in a different but related anti-colonial context might call 'fugitivity' (Harney and Moten, 2013) does not lessen my desire for clarity, and neither do I have recourse to lyricism as some alternative to theorisation. The impossible goal is to fashion a style that is at once 'adrift and moored', allowing for a theoretically articulate but necessarily incomplete communication of an idea; one that is critical, clinical and creative at the same time, open to the forces that inhabit the other side of the skin, in the im/material space of the *oikos*. In opposition to the melancholy of speculative realism or the celebrations of agential realism, the *ecological realism* that interests me in this book – the thing I hope to evoke – is neither optimistic nor pessimistic. Quite simply, it is a way of being receptive, attuned to the potentials of the 'not yet' and 'other than' at play in any theatrical event.

There is no appeal to efficacy beyond the event in my idea of theatre ecology. No desire to militate for action in the so-called, 'real world'. This is because, as I show in my reading of Karen Barad in Chapter 3 (see pp. 106–10), there is no such thing as the 'real world' – at least not a world that is there to discover as an object consubstantial with itself, hermetically sealed in on itself. As I contend again and again in this book, theatre already *is* action. Its ecology is found in how it unleashes the becoming of the non-human, both in the affect-strewn thoughts it gives rise to and in the writing it provokes. There is, then, little to be gained for theatre in representing ecology as mere theme or figure, precisely because, as I show in Chapter 2, ecology is theatrical, a spatio-temporal process in which all matter is haunted by the virtual impulse that the earth has to be other than itself, here *and* elsewhere – 'doubled', as Artaud might put it.

Conclusion

In this chapter, I have tried to do two things: first, to map a contested field of ecological criticism in Theatre and Performance Studies by focusing on the differences inherent in the dominant methodological approaches advanced by Chaudhuri, Marranca, Kershaw and Read; and, second, to provide a critical inventory of subsequent scholarship that has drawn on these models, explicitly and implicitly, by focusing on what they both afford and deny. I finished by gesturing towards the possibility of another, more medium-specific method of 'doing' theatre ecology that concentrates on the immanent ecology of the theatrical event. The rest of this book is an attempt to unfold the idea inherent in that dual premise, explaining, in the process, what it may mean for Theatre and Performance Studies to *theatricalise* ecology and to ecologise theatre. The ramifications of the double thinking are extensive. For to propose an idea of theatre ecology is not just to intervene into a genre, to give examples of performance that can be read environmentally, it is to rethink what theatre and theatricality have been considered to be, historically, aesthetically and conceptually.

Notes

1 Luhmann's work has often been criticised for accepting capitalism and liberalism as systems to regulate and improve, rather than as ideologies to overthrow and abandon. In this respect, Luhmann forms part of what the Tiqqun Collective critique as the 'Cybernetic Hypothesis', a mode of neoliberal governance constructed around managing the flows of data and information. The point is always to preserve the system, to engineer change so that everything remains the same. As Tiqqun put it: 'The cybernetic hypothesis thus expresses no more or less than the "end of politics"' (Tiqqun, 2010: 8).
2 I use the word 'humanist' here, since much environmental thought still invests in ideas of intentionality, volition and self-presence, at the very moment that it argues for the need to do away with anthropocentrism. As will be explained further in Chapter 2, I do not intend to do away with humanism in my idea of a theatre ecology; rather, I argue for a posthumanism, in which some humanist values – the healthiest ones – are retained: equality, democracy, the right to live well. But at all times, I am suspicious of ecological thought that continues to place the onus on consciousness, self-presence and simplistic understandings of cause and effect. So while I am aware that the *anthropos* can never be abandoned, its centrality and exceptionalism most certainly can be.
3 The term was advanced by Paul J. Crutzen and Eugene Stoermer in an article in 2000 to designate a new stratigraphic era in the earth's natural history, the geological era of man (*anthropos*) (Crutzen and Stoermer, 2000).
4 I discuss the longstanding relationship between anthropocentrism, humanism and the history of the western stage in Chapters 4 and 5.
5 For more on this topic in Theatre and Performance Studies, see the special issue of *Nordic Theatre Studies* 'Theatre and the Anthropocene' edited by Stephen E. Wilmer and Karen Vedel (2020), and my essay 'Thinking Materially: Verdonck in the Anthropocene' (Lavery, 2020).
6 Stengers's notion of the pharmacological draws on Jacques Derrida's usage of the term in his 1968 essay 'Plato's Pharmacy'. In Derrida's text, the *pharmakon* is associated with the 'poisonous cure' of writing, the ambivalent thing that troubles self-presence and yet is indispensable for self-knowledge and the proper transmission of ideas (Derrida, 1981: 69–70). The *pharmakon* is a figure for undecidability, doubleness and the impossibility of wholeness and totality.
7 While Lisa Woynarski offers a very useful survey of the historical development of ecocriticism in Theatre and Performance Studies, she is reluctant to subject the critical 'waves' she discusses to any real critique (Woynarski, 2020: 14–18). The same criticism could be levelled at Julie Hudson who, although willing to take issue with Chaudhuri's critique of naturalism as being too human-centred, does not subject the thinking of other scholars, or indeed the idea of ecology, to the same level of scrutiny (Hudson, 2020: 15–20).
8 At the start of the 2010s, Wallace Heim (2012a) and Dee Heddon and Sally Mackey (2012), in two separate publications, urged theatre scholars to start engaging seriously and copiously with issues of ecology and environment.

The desire was for 'more work' (Heim, 2012a: 215) and better theorising. While much in that dual interpellation has been met, the question of critique, which, as I explain in the main body of this chapter, is invariably bound up with methodological issues, has remained largely unexplored.

9 One wonders, in light of an essay by Michael Marder (2018: 141), if the prefix 'eco' is the best term to use in these attempts to specify where theatre's contribution to ecology may reside. The desire to be economic in one's usage of language obscures an often violent relationship inherent in the very etymology of the word ecology itself, and which pertains to the suffix *logos*, meaning law, economics, measure. From this point of view, the 'eco' in ecology is indicative of dubious economic logic, a desire to circumvent complexity in order for the free circulation of ideas to occur in a frictionless space of academic 'exchange'.

10 Hudson, perhaps, gets closest to understanding how theatre can ecologise when she defines ecotheatre as 'not limited to the idea of productions designed to deliver a campaigning message'. Instead, she states that 'it positions theatrical events as ecosystemic, therefore also unpredictable and potentially equivocal in meaning and affect' (Hudson, 2020: 4). For her, theatre – potentially all theatre – is ecological because it functions like an ecosystem. While I have enormous sympathy for this point of view, it still assumes that theatre's ecology is reflective; its mode of operation a mirror for how ecosystems function. By contrast, I am more interested in how ecology itself is theatrical, a reversal which moves the centre of gravity from metaphor to real, from analogy to homology, a kind of common fold. It is also telling that the majority of Hudson's examples are performances that deal quite explicity with environmental questions. They do not really 'stretch' audiences or leave them confounded about the content on offer.

11 Many ecocritics, for instance, do not even mention *Staging Place* but prefer to concentrate on Chaudhuri's shorter 1994 text '"There Must Be a Lot of Fish in that Lake": Toward an Ecological Theater' (Chaudhuri, 1994). The same applies to the critical reception of Kershaw's monograph. Somewhat oddly, the most interesting appraisals of *Theatre Ecology* are found in review articles. The long review by the environmentalist and playwright Paul Brown in the Australian journal *Performance Paradigm* (Brown, 2009) remains unsurpassed for its understanding and critical rigour.

12 The consequences of that 'obscuring' are inherent in the first three texts to be published in Palgrave's excellent *Performing Landscape* series. Not only is Marranca's name absent from them, but her particular approach to theatre ecology as a biocentric practice, an environmental form in itself, is occluded, in favour of other, more normative readings of landscape and Nature, rooted in representationalism and/or site-based interventions.

13 By focusing on specific monographs, I am aware of overlooking the totality of what these innovative scholars have contributed to the development of the field of theatre and ecology. There is no specific place, for instance, for Chaudhuri's, Kershaw's and Read's substantial work on expanding the performance collective to include animals, insects, grasses and plants; and neither is there the space to discuss, in any detail, Kershaw's and Chaudhuri's practice-based investigations

into climate change, landscape and the Anthropocene. For more select publications on these topics, see Chaudhuri's (with Shonni Enelow) *Research Theatre: Climate Change and the Ecocide Project* (Chaudhuri, 2013), and *The Stage Lives of Animals: Zoesis and Performance* (Chaudhuri, 2016b); Kershaw's essays 'Dancing with Monkeys? On Performance Commons and Scientific Experiments' (Kershaw, 2012a); '"This is the way the world ends, not ...?": On Performance, Compulsion and Climate Change' (Kershaw, 2012b); and 'Projecting Climate Scenarios, Landscaping Nature, And Knowing Performance: On Being Performed by Ecology' (Kershaw, 2016); and Alan Read's monographs *Theatre & Everyday Life: An Ethics of Performance* (Read, 1994) and *The Dark Theatre: A Book about Loss* (Read, 2020). However, what I will say, as a sort of recompense, is that their methods remain remarkably consistent across their respective publications, for the most part. The major change, perhaps, is tonal, a drift that pertains to a darkening of the mood, especially with respect to Chaudhuri's and Read's thought.

14 For more on Gibson's notion of affordances and ecological perception, in general, see *The Ecological Approach to Visual Perception* (1979). Although theatre is a primarily visual medium, the affordances that interest me in this book go beyond the eye and respond to other sensory possibilities and allowances. The point to grasp in this analogy with Gibson is that theatre does not represent; it is a milieu that produces.

15 Joanne Tompkins has celebrated the field-defining nature of *Staging Place* in the 'Re-Readings' section of the journal *Contemporary Theatre Review*. However, she makes no mention of how central the text is to the development of a specifically dramatic notion of ecocriticism (Tompkins, 2020: 408–9).

16 While Chaudhuri is acutely aware of the bad faith pervading Old Ekdal's loft, she is nevertheless loath to dismiss it, entirely. In keeping with her symptomology of performance, she argues that its presence allows for the violence of modernity to be perceived. The loft is indicative of what one could call modern drama's 'negative ecology', the sense in which, in Ibsen's work, in particular, the loss of contact with 'Nature' is registered obliquely, in the spatial behaviour of characters that are disorientated, caught between two worlds, discomfited. Where Nora's geopathology in *A Doll's House* produces forgetfulness through flight, Old Ekdal's remains in contact with geopathology's wound: 'From the beginning of the play, Old Ekdal is represented as the one who is out of place [...]. It makes him, in fact, the exemplar of the most radical geopathic experience, one beside which those of Gregers, Hjalmar, and even Nora, in her play, seems pale and trivial. The displacement Old Ekdal suffers from is neither temporary nor curable; it is the loss of nature itself. This loss lies beyond ritual recovery, but not beyond consequence' (Chaudhuri, 1995: 75–6).

17 For more on the relation between Absurdism and planetary extinction, see my entry on 'Late Modernism' in John F. Deeney and Maggie Gale (2016a: 545–67) and also all of the essays in Carl Lavery's and Clare Finburgh-Delijani's 2015 collection *Rethinking the Theatre of the Absurd: Ecology, Environment, and the Greening of the Modern Stage* (Lavery and Finburgh-Delijani, 2015).

18 It is also more normative, reflecting a reassuring structure of thesis, antithesis and synthesis.
19 In his celebrated account of the Bonaventure Hotel in *Postmodernism, or The Cultural Logic of Postmodernism*, Jameson famously equates postmodernism with a loss of the real (Jameson, 1992: 39–44).
20 It sometimes seems in Chaudhuri's argument that modern western drama had to go through a kind of Hegelian process of alienation from 'Nature' in order to locate a better, more progressive ecotheatre in the postmodern, multicultural stages of the US in the mid-1990s. In light of the idea of virtual ecology in the work of Deleuze and Guattari and also Karen Barad's quantum queering of time, such linearity merits critique today. For more on Barad's spectral time ecology, see pp. 106–10 in Chapter 3 of this book
21 Hudson is one of the few critics to concentrate on Chaudhuri's reading of Gray's work, but she does so from the point of view of 'frugal' storytelling (Hudson, 2020: 130–50). Conversely, I look to stress the emphasis that Chaudhuri puts on spatiality in her analysis of Gray, the sense in which the minimalism of the stage produces a phenomenology of togetherness.
22 I am aware of course that contemporary screenings of films can also produce this site-based sensibility, but the fact remains that the actors and actants are not present in the same way. The cinematic event, in other words, is still predicated on looking at images not in being together with fleshed bodies.
23 Oliver provides a pithy account of her argument in the 2018 essay 'Earth: Love It or Leave It', when she explains how the God's-eye view of the earth from space 'makes us both want to protect our vulnerable and fragile planet and to escape from this insignificant speck' (Oliver, 2018: 347). Whatever option is taken, the human remains outside the earth, the one who looks on, seeing only false unity.
24 At roughly the same time as Marranca was formulating her ideas on landscape, theatre and ecology, Elinor Fuchs was developing a similar model of ecotheatre in her much-cited essay 'Another Version of Pastoral', initially published in the 1994 special issue of the journal *Theater* (Fuchs, 1994).
25 Marranca's thinking of theatre ecology grew organically from her 1977 publication *The Theatre of Images*, which looked at the work of Robert Wilson, Mabou Mines and Richard Foreman, For a different but related view on the ecological image, see my essays on Philippe Quesne, and Mike Brookes and Rosa Casado (Lavery, 2013, 2019a).
26 Unlike Kershaw, however, Marranca does not follow this double articulation of theatre ecology up in any detailed way. Her focus, as she states, is 'not concerned with sociological analysis or political economies', but with a 'lyrical, philosophical, and experimental approach to arts practice and the writing about it' (Marranca, 1996: xv).
27 The necessity to foreground 'Nature' is evident in this passage from Plumwood: 'To be defined as "nature" in this context is to be defined as passive, as non-agent and non-subject, as the "environment" or invisible background conditions against which the "foreground" achievements of reason or culture (provided

typically by the white, western, male expert or entrepreneur) take place. It is to be defined as a *terra nullius*, a resource empty of its own purposes or meanings, and hence available to be annexed for the purposes of those supposedly identified with reason or intellect, and to be conceived and moulded in relation to these purposes' (Plumwood, 1993: 4).

28 Autopoesis refers to the cybernetic operation whereby the system gives birth to itself, evolving according to its own recursions and relations.

29 For more on the double bind, see Bateson's cybernetic theory of alcoholism (Bateson, 1972: 307–37).

30 McKenzie's innovative and stylish text was the first to break ranks with the political 'hype' accredited to performance and performativity by theatre and performance scholars. Whereas performance was routinely championed as a vehicle for political emancipation, McKenzie highlighted the relationship between performance, capitalist managerialism, information technology and control societies. See his 'Introduction: Challenges' for a good overview to the remit of his book (McKenzie, 2001: 3–26).

31 When it comes to responding to humankind's derangement of the environment, anthropocentric solutions, no matter how well-intentioned, 'only escalate' the initial problem, Bateson claims, by putting 'positive feedback' into the system, which in turn produces 'schismogenesis', a tipping point whereby all hope and balance are lost (Bateson, 1972: 323–4). In its pathological desire to act, the human subject once again positions itself as the central agent, a misplaced investment in sovereignty that, in keeping with the logic of paradox and positive feedback, is the very thing that will destroy it, its tragic flaw. Hence the reason why 'weakness' is a strength, for Bateson and Kershaw: it creates negative feedback.

32 Kershaw is more generous about Fuchs's notion of the pastoral, seeing it as offering 'a more embedded take, so to speak, on the relationship between ecology and aesthetics' (Kershaw, 2007: 309). In making this claim, however, Kershaw seems to gloss over the fact that Fuchs's pastoralism is still concerned with looking at things.

33 In more recent years, Kershaw has preferred to overcome what he sees as the 'double-bind' of black box theatre by investigating how the world 'performs on us' through a series of practice-based experiments.

34 Comparing the traditional theatre space to the vast glass domes of Biosphere II in Arizona and the Eden Project in Cornwall, Kershaw contends that ecological performances taking place indoors only perpetuates the pathologies they were intended to avoid: '[W]e might view Biosphere II as an attempt at restitution, an act of healing aimed to make amends with "nature" […]. If so, the hermeticism of the glass ark is fabulously ironic; as in theatre the more one yearns for contact – with "nature" or its substitutes – the more out of reach it will become. Hence, paradoxically, the ecology of twentieth-century theatre in the West, like Biosphere II, has reproduced the environmental pathologies that an ecologically aware theatre might most wish to avoid' (Kershaw, 2007: 316).

35 I am using the term environmental here in the way that Richard Schechner (1985) and Arnold Aronson (1981) use it. Environmental theatre is not about the environment in any ecological sense; rather, it refers to an immersive, participatory mode of staging.

36 Paul Brown makes the insightful comment that Kershaw is not so much interested in environmental theatre as in 'applying ideas of ecology to the study and analysis of theatre and performance' as ecosystems in their own right (Brown, 2009: 121). Similarly Jenny Hughes in an equally perceptive review in *New Theatre Quarterly* is correct to say that what interests Kershaw is the 'wholesale adoption of radical ecology as a theoretical frame' (Hughes, 2009: 103).

37 Kershaw's use of the concept of the 'black hole' resonates, closely, with Deleuze and Guattari's formulation of the term, as a kind of loss or destruction that even politically progressive movements can find themselves getting sucked into. For more on the relationship between Fascism and the black hole, see Deleuze's and Guattari's *A Thousand Plateaus* (Deleuze and Guattari, 1987: 228).

38 Transversality is a complex term, and one that has many different applications in Guattari's thought, depending on its context. Gary Genosko provides a useful and succinct definition in his essay 'The Life and Work of Félix Guattari: From Transversality to Ecosophy' (2000), when he says that 'Transversality is a tool to open hitherto closed logic and hierarchies' (Genosko in Guattari, 2000: 119). Otherwise put, transversality is the art of bringing disparate things together, a travelling across borders.

39 Although Kershaw draws enthusiastically from Guattari, he pays little attention to the strong aesthetic dimension in the latter's thinking about 'virtual ecology' (Guattari, 1995: 88–97). My book, as I discuss in *Polemos* and Chapter 3, is an attempt to use Guattari's ideas, differently, to pay attention to how the autonomous artworks express their own, immanent ecology.

40 Kershaw's own double bind centres on his attachment to ideas of equilibrium and balance. In privileging these notions, he runs the cybernetic risk of keeping the status quo as it is, of sustaining the system, not breaking with it (Kershaw, 2007: 248). This is a critique that has been levelled at Batesonian theories of information and indeed cybernetics in general, as I pointed out in note 1 in this chapter. It also departs, significantly, from Guattari's thinking on machines, which he values precisely because they break down.

41 In both Dominic Johnson's (2008: 514–15) and Joe Kelleher's (2008: 181–3) reviews of Read's *Theatre, Intimacy & Engagement*, the word ecology is notable by its absence. Likewise in Martin Harries's rereading of *Theatre and Everyday Life* in the 'Double Take' Section of *Theatre Research International*, no mention is made of Read's early concern with the environment (Harries, 2020: 362–4). It is also often forgotten that Read is mentioned in Chaudhuri's 1994 article on ecology and that he contributed an important essay to Gabriella Giannachi's and Nigel Stewart's 2006 collection *Performing Nature: Explorations in Ecology and the Arts*.

42 Derrida makes a similar point in his reading of the human, the impossible animal, in *The Animal that Therefore I Am* (2008).

43 This is close to Hans-Thies Lehmann who says that theatre is an act of arriving, not arrival, 'presencing' not presence, an aesthetics of *'startling'* (Lehmann, 2006: 143; original italics).
44 It is useful to compare Read's argument here with Ralph Yarrow's in 'Writing Degree Zero' (2001). Read's lessness is very close to Yarrow's zero, on account of the fact that one gets to the outside by going through theatre.
45 Agamben is drawing on Aristotle's usage of these terms. In Aristotelian philosophy, the human is a special creature, since it is made up of two forms of life: *bios*, which he associates with thinking, culture and identity (having a name) and *zoē*, which is creaturely, biological, anonymous being. Agamben contends that reducing people to bare life is the very thing that allows governments and ideologies to deny their humanity, to make them superfluous and thus destined to be killed with impunity. Read's take on *zoē* is very different. For him, *zoē* is ecological life, the life that humans have historically refused.
46 For all his concern with paradox and failure, Kershaw, curiously, pays little or no attention to the ecological potential inherent in theatre work that deconstructs itself in his way.
47 An exception is found in Stephen J. Bottoms and Matthew Goulish's collection *Small Acts of Repair: Performance, Ecology and Goat Island*, although even here the discussion is somewhat truncated (Bottoms and Goulish, 2007: 34–5).
48 The emergence of ecology as a key topic of our times has resulted in an explosion of playwrights and companies wanting to engage with it directly and in different forms. In Anglophone practices, Katie Mitchell, and Sheila Ghelani and Sue Palmer have experimented with documentary dramas and lecture performances; Stan's Café with low-carbon performance on bikes and site-specific installation; Simon McBurney and Selina Thompson with solo performance and storytelling; Chantal Bilodeau, Amy Berryman, Caridad Svich, Ella Hickson, Alasdair McDowell, Anne Washburn, Caryl Churchill with dramatic plays and readings; Fevered Sleep with video and multi-media works; and Platform with site-based protest performance and audioworks. These are only a few examples from a vast number of environmentally committed practices.
49 Churchill's 1989 play *Ice Cream* also features in Chaudhuri's *Staging Place*, but it does so in terms of a spatial politics of tourism, rather than any explicitly ecological politics (although when it comes to questions of dwelling, the two, of course, cannot be separated, ultimately).
50 Downing Cless's work is the exception. Throughout his text, he draws on his own experience as a director of the works he writes on.
51 Given his interest in posthuman assemblages, it is ironic that Bruno Latour's lecture performances (*Cosmocoloss. A Global Climate Tragic Comedy* (2011), *Gaia Global Circus* (2013) and *Inside* (2017)) should place such primacy on language as the primary conduit for ecological ideas. In Latour's theatre, and also in Katie Mitchell's work in this area, the medium is treated as a transparent thing, something that, though sensuous, disseminates information in a relatively clear manner. There is no sense that it is a network or actant full of slippages, viscosities, random occurrences, misrecognitions, affective dissonances.

The autonomy seems tempered, lying always with the sovereign human subject. For more on contradiction in Latour, see Chapter 2, p. 82 in this book.

52 In her analysis of the play *Grasses of a Thousand Colours*, Una Chaudhuri has argued for the need to experiment with new forms of dramatic incommensurability to reflect the derangement of the Anthropocene. Somewhat strangely, however, she relies on a largely intellectualist notion of theatre – a kind of Platonic thesis play – to do that. Instead, then, of returning to the cave, of acknowledging bodily entanglement and interdependence, becoming sensate, Chaudhuri wagers yet again on the *logos* of the sky god, on the transcendent capacity of thought to extricate humanity from the very mess that thought has caused. For more on Platonic anti-theatricality and ecology, see pp. 122–8 in Chapter 4 in this book.

53 In this activist tradition, site, understood in its most expanded sense, as bodies, institutions and actual places, is crucial for both witnessing and staging environmental and ecopolitical interventions. In terms of Theatre and Performance Studies, one thinks of the substantial contributions to environmentalist notions of site in the work of Sian Rees (2018), Denise Varney (2018), Peta Tait (2018), Sarah Ann Standing (2019), Mojisola Adeybayo (2021 ongoing), and Isabelle Frémeaux and John Jordan (2021). Other criticism worth mentioning in this context is work that owes its inspiration to the ideas of Grant Kester, Nicolas Bourriaud and Claire Bishop, all of whom are closely connected with participatory art practices. Useful texts here include Daro Montag's *Artful Ecologies* (Montag, 2008), a collection of papers from the Art, Nature and Environment Conference held in Falmouth in 2006, in which a number of artists and theorists, including Suzi Gablik, and Tim Collins and Reiko Goto, reflect on work made in direct response to the climate crisis; Malcolm Miles's *Eco-Aesthetics: Art, Literature and Architecture in a Period of Climate Change* (Miles, 2014), which despite its title, reflects on the politics of site in ecoperformance practices by artists such as Mierle Laderman Ukeles, Cornford & Cross, and Heather and Ivan Morrison; Heather Davis and Etienne Turpin's *Anthropocene Encounters: Encounters among Aesthetics, Politics, Environments and Epistemologies* (2015), a text that brings together artists and leading theorists to think through what a viable ecopolitical art practice for the Anthropocene might look like; and T. J. Demos's engagement with art and activist practices in Mexico City, Canada, Tanzania and India in *Decolonizing Nature: Contemporary Art and the Politics of Ecology* (2016).

54 The ethos of much practice-based ecoscholarship is to embody and enact discursive ideas from other fields and disciplines. The work of new materialist and posthumanist thinkers, such as Karen Barad, Jane Bennett and Bruno Latour has proved particularly popular in this regard. In the best of this work, performance transforms ecocritical thinking and vice versa by opening up different actants and publics, embracing corporeal modes of interventions and bringing together concepts and ways of being that might appear incompatible from a theoretical standpoint. For key works that deal with vibrant matter in performance, see Paula Kramer (2012), Chaudhuri and Enelow (2013), Minty Donald

(2014, 2019), Minty Donald and Nick Millar (2014–16), Tanja Beer (2016), Kershaw (2016), Stephen Scott-Bottoms (2019) and Vincent Roumagnac (2020).

55 Critics who take an intersectional approach to the ecology of site, include Pavithra Vasudevan (2012), Elizabeth Stephens and Annie Sprinkle (2012), Lisa Woynarski (2015, 2020), Bronywn Preece (2018), Helena Grehan (2018) and Louise Ann Wilson (2019).

56 Although my essay 'Mourning Walk and Pedestrian Performance: History, Aesthetics and Ethics' is not about ecology *per se*; it suffers from a decidedly enchanted – and thus humanist – notion of landscape (Lavery, 2009: 48–52). The idea is to use the environment to bolster the subject, not to disrupt its agency or to question its humanness.

57 Ferdinand's search for a 'worldly ecology' is committed to overcoming 'the fracture [that] separates the colonial history of the world from its environmental history'. According to Ferdinand: 'This [fracture] can be seen in the divide between environmental and ecological movements, on the one hand, and postcolonial and antiracist movements, on the other, where both express themselves in the streets and in the universities without speaking to each other. This fracture is also revealed on a daily basis by the striking absence of Blacks and other people of color in the arenas of environmental discourse production, as well as in the theoretical tools used to conceptualize the ecological crisis' (Ferdinand, 2022: 3).

58 I use 'abstract' in the technical way an art historian would, to designate a mode of mimesis that abandons reflection and resemblance and focuses instead on the decidedly materialist existence of colours, shapes and textures. One could easily replace abstract with immanence, in my thinking.

59 There is a real sense that Kershaw's ideas could be used for similar ends, but his work on cybernetics is never applied to actual theatre events. Although Hudson cites the influence of Kershaw on her thinking (Hudson, 2020: 2), her ecosystemic approach to performance is actually much closer to Marranca's method – a correspondence she duly acknowledges (Hudson, 2020: 23).

60 In *Theatre Ecology* Kershaw, too, touches, all too briefly, on the non-human quality of performance in a chapter on Buster Keaton, Stelarc, and Coco Fusco and Guillermo Gómez-Peña (Kershaw, 2007: 227–38). For him, the non-human appears to be associated with a common identity 'shared by human and non-human organisms' (Kershaw, 2007: 237). By contrast, I approach the non-human as a pre-individual, inorganic flux that combines energy and matter. Instead of being cellular, it is virtual. It is more than biological.

61 For a very different, more ecologically attuned practice of theatre in auditoria, see Martin Welton's work on atmospheres and atmospherics (Welton, 2012, 2018), and also his writings with Adam Alston on theatre in the dark (Alston and Welton, 2017).

62 This transgression of the life–death boundary is not only because theatre deals with ghosts, but also because it shows human beings in the very act of dying itself, in consuming themselves. Consider this statement from Tim Etchells: 'Death haunts all performance […] In my own work with Forced Entertainment

I'm struck by the fear that the performers are publicly rehearsing their own deaths, plotting lives for their own dead selves' (in Heathfield, Templeton and Quick, 1997: n.p.).

63 In the scene from *Our Mutual Friend* that Read alights on, the villain Riderhood is fished from the Thames and an attempted resurrection takes place in a nearby tavern. As they watch Riderhood pass away, the cast of characters, who had previously detested Riderhood, suddenly take a keen interest in his survival. Their passion, however, is not for the individual but for life itself (*zoē*).
64 I write im-mediate with a hyphen in order to trouble the so-called temporal presence and spatial completeness habitually associated with immediacy. The same holds for my spelling of inter-ested. The hyphen underlines the fact that to be interested is to be placed between things (essences).
65 For more on how the eye 'breathes' in the matter of performance, see my essay on the ecological image in the work of Lee Hassall (Lavery, 2019b).
66 One thinks, for instance, of how humanism's project is always about overcoming difference, of finding a sense of oneness, doing away with the dangerous supplement that it needs to give itself the semblance of wholeness. For more on this, see my discussion of Claire Colebrook on pp. 81–3 in Chapter 2 in this book.
67 This is how Cohen and Duckert define veering: 'Far from merely environing the human in anthropocentric ways – Michel Serres's worry about the term *environment* – *Veer Ecology* acknowledges a world of inhuman forces, dynamic matter, and story-filled life that inevitably go off course. They act, they drift, they swerve, and resist. In diverging from human domination, they disrupt secure dwelling in ways that are catastrophic, pleasurable, orbit changing. Besides a swift change of subject or direction, *veer* describes wind's swirling motion. Though not the world's only sudden element, air well conveys the dynamism embedded in *veer*, the propensity it designates to circle back, to whirl as a vortex' (Cohen and Duckert, 2017: 3). My only caveat with their beautiful definition is the onus they place on story.
68 For more on the difference between method and practice, see Karin Murris's and Vivienne Bozalek's article 'Diffraction and Response-able Reading of Texts: The Relational Ontologies of Barad and Deleuze' (Murris and Bozalek, 2019: 1–3).

2

Theory (theatricalising ecology)

Introduction

Towards the end of Chapter 1, I proposed a new idea for a theatre ecology. In Chapter 2, I look to clarify what the concept of ecology refers to in that idea. I am obliged to move beyond disciplinary boundaries in this chapter, since the posthumanist ecology that my idea is based on both borrows and departs from extant theories and debates about the meaning of posthumanism in the Environmental Humanities. It cannot be thought in isolation. However, notwithstanding its interdisciplinary nature, it is important to insist on the singularity of theatre ecology. Where the major practices of posthuman ecology in the Environmental Humanities have tended to place the non-human beyond the human and to approach it as an outside that needs to be embraced – a machine, an animality, an earth – I contend that the non-human is lodged in the very heart of the human, a dissonance it is unable to master or still. Similarly, where much posthuman ecocriticism, in tandem with ecocriticism in general, invests in narrative as the form best suited to create new relations with the non-human world, I look to theatre as offering an alternative method; one where ecology operates through bodies, not words, and where sensation is a more powerful stimulant to change than interpretation. Everything is to be gained in this 'theatricalisation of ecology', since, as my reading of Gilbert Simondon explains in the final section of this chapter, ecology itself operates according to theatrical principles of transmutation and transformation. In that respect, the chapter looks to make two related contributions. On the one hand, to set forth an immanent theory of ecology for Theatre Studies, rooted in a particular understanding of posthumanism; and, on the other, to intervene into aesthetic and philosophical debates in the Environmental Humanities at large by theatricalising the very notion of ecology itself and, by doing so, to encourage critics to question their allegiance to narrative as the most effective way of communicating ecological experience.

Meanings of ecology

Although the etymology of ecology is easy to determine, coming, as it does from the Greek *oikos* (home) and *logos* (law, measure), this does little to explain its varied and often competing usages in the Environmental Humanities. In the disparate disciplines that make up that paradigm, ecology no longer designates a simple biological relationship with one's home, habitat or *Umwelt*, as one sees in the original scientific usage of the term by the German zoologist Ernest Haeckel in the late 1800s.[1] Rather, it is a complex term, a 'keyword' in Raymond Williams's lexicon (Williams, 1988: 111).[2] Ecology's inbuilt dissensus is evidenced, in the most empirical of ways, by simply listing some of the labels that are now attached to it in the Environmental Humanities: ecophenomenology, environmental justice, posthuman ecology, ecofeminism, ecodeconstruction, political ecology, elemental ecology, queer ecology, green ecology, blue ecology, dark ecology, black ecology, Indigenous ecologies etc. As these examples demonstrate, ecology is overdetermined, an unstable, plastic term whose meaning or function is dependent upon the standpoint it is approached from. Where some, for instance, see ecology as an attempt to rediscover more embodied relations with 'Nature'; and some as a way to contest how 'Nature' has been mobilised as a signifier; there are others who approach it, more radically, as a specific way of thinking and being, an onto-epistemology that takes it distances from the western humanist model of volition, consciousness and agency.

Given these multiple meanings and diverse applications, it is imperative that the ecology in theatre ecology is precisely defined. Only then will it be possible to discern the specific type of contribution that a minor art such as theatre can make to the production of a worthwhile ecological *habitus*. More than academic accuracy is at stake. Forgetting to interrogate and define the word ecology runs the risk of complacency, missing the 'cut' that matters. The last thing one wants is to perpetuate a culture of perverse performatives, rhetorical interpellations that deteriorate an already bad scenario, precisely because they are so confident about their capacity to change the environmental status quo for the better. At a time of empty rhetoric, sanctimonious sentiment and endless summit meetings, something else is required to transform relations with 'Nature', a new way of operating. To get to grips, then, with what the troublesome word 'ecology' in theatre ecology actually means in this book, it is useful to trace the mutations of the concept in the Environmental Humanities as it has played out over the past few decades. The intention is not simply to provide a frame of reference; it is to express critical and political allegiances. To return to Malcom Ferdinand's point about decolonial ecology, there is nothing so humanly relevant, so politically charged, as the environment (see p. 57 in this book).

As Greg Garrard (2012) and Ken Hiltner (2015) have explained in depth, the status of the word ecology has unfolded in three waves, since it first appeared in literary criticism in the 1960s and early 1970s. In the first wave, associated with Romanticism, deep ecology and early ecofeminism, the aim was to return to 'Nature' as a universal and intrinsic good, to look for a proper equilibrium with the environment beyond the deranging distortions of patriarchal thinking and exploitations of capitalist economics.[3] In the second wave, which approximately runs from the 1980s through to the 1990s, the contradictions inherent in these essentialist notions of 'Nature' and balance were highlighted and numerous thinkers and activists looked, instead, to show the discriminatory violence at play in problematic concepts such as 'Nature', 'wilderness', 'evolution' etc. Importantly, in these years, there was also a new onus in moving ecology away from a phenomenology of environmental experience and embracing, instead, the work of theorists and cyberneticians such as Niklas Luhmann, Humberto Maturana and Gregory Bateson, all of whom saw ecology as both autopoetic network and structure, a 'more than human' system with no apparent outside.[4] In Bateson's 'ecology of mind', for instance, ecological thinking, as I touched on in the previous chapter (see pp. 42–4), is not confined to human consciousness alone, but on the contrary is 'expanded' to include all the information, signals and messages transmitted by diverse organisms in a common milieu. Bateson's key insight, influenced, as it was, by the findings of Maturana on the immanent creativity or emergence of the system, is that the observer does not stand outside of the circuit they are looking at, as proposed in the early cybernetic work of Norbert Weiner in the 1940s and 1950s; rather, they are part of it, affecting and being affected by the feedback loops that their presence in the system invariably provokes. Notwithstanding a traditional environmentalist concern with 'balance', Bateson's ecology of mind allows for a new kind of ecology to appear, one that is provoked by the autopoesis of the system itself and which humans have to regulate from the inside, a situation which, as I explained in relation to Kershaw's method in Chapter 1, complicates the agency, autonomy and exceptionalism promoted by Enlightenment notions of humanism, based on reason, autonomy and self-knowledge.

Many of the developments that characterised second-wave ecology are still in operation in today's posthumanist third wave, but with three major differences. First, that the politics of discourse have been supplemented with a politics of bodies, sensation and matter – a politicised 'new materialism' that is aware of how global majority subjects and Indigenous populations create worlds that contest the destructive credo of western humanism; second, that ideas of interactivity (relations between) have been challenged by those of intra-activity (relations within), a shift that challenges deep-rooted attachments to bounded and substantialist notions of identity in the name of what Mel Y. Chen terms 'queer intimacy' (Chen, 2011: 278);[5] and, third, that the cybernetic faith in maintaining equilibrium and balance has run its course, with chaos and complexity now being regarded as forces to embrace, breakdowns that may lead to breakthroughs.[6] Confronted with a new climatic regime that is experienced as an everyday reality, third-wave ecology is no longer a minority pursuit, confined to a largely white, middle-class elite concerned with overcoming their alienation from the so-called natural world. It impacts on everyone on the planet and demands radical changes in politics, economics, ethics.

More than that, it argues for a fundamental transformation in how humans think of themselves as both a political and planetary species, no matter how difficult that is. In contemporary posthumanist understandings of ecology, as Dipesh Chakrabarty argued in his pioneering article 'The Climate of History: Four Theses' (2009), it is senseless to divorce history from natural history in any definitive sense. 'The wall between human and natural history has been breached' (Chakrabarty, 2009: 221). As a result, everything is filtered and shot through with everything else, to the point that the very notion of a natural disaster is now out of the question. To exist as 'a species being' is to recognise oneself as a geological agent, a terrestrial creature:[7]

> The anxiety global warming gives rise to is reminiscent of the days when many feared a global nuclear war. But there is a very important difference. A nuclear war would have been a conscious decision on the part of the powers that be. Climate change is an unintended consequence of human actions, and shows, only through scientific analysis, the effects of our actions as a species. (Chakrabarty, 2009: 221)

In this 'equivalence of catastrophes', the human is neither central to nor displaced from ecological debates, as it tended to be in the first two waves (Nancy, 2014: 3).[8] Now, the subject is interrogated, naturalised and earthed; its pretensions for transcendence more comprehensively debunked, its philosophical systems more radically deconstructed. And yet while traditional humanism has been most certainly disavowed in this third wave, the human or *anthropos* continues to be stubbornly present within it as the limit that can never quite be surpassed, a persistence that demands vigilance.

Claire Colebrook offers some germane insights into why such vigilance is necessary when she cautions against aggressive forms of humanism that still remain latent and unquestioned within certain forms of posthumanist philosophy. Ironically, for Colebrook, posthumanist thinking is often predicated on a repressed humanism that undermines the environmental shift it looks to bring about. This is because the goal of the humanist subject for Colebrook was not so much to depart from 'Nature', as Bruno Latour has argued so extensively in *We Have Never Been Modern* (1993), but rather to find a way of becoming one with it, annulling its separateness, perfecting its essence through the incorporation of otherness.[9] As Colebrook has it, humanism is a philosophy of erasure, and what it wants to erase is the tension that constitutes its very being. To be posthuman, for Colebrook, is not about becoming animal or machine, and it is certainly not about returning to a pristine natural world that 'might have remained as such had "we" not been so world-disturbing' (Colebrook, 2015: 225). Rather, it is about learning to exist as an im-proper creature, one that is deprived of unity and troubled by an alterity that is always already at work within in it:

> [T]he human is not a being or event within time, but a structure of erasing what has come to appear as human, aiming to be nothing other than an event of erasing its own contaminating and inauthentic past. [...] The human [is] always trying to efface itself in order to find what it is not itself and become at one with a life or being that never can be admitted as radically distant and difficult. (Colebrook, 2015: 224)

The philosophical and environmental stakes of Colebrook's perspective are far-reaching. Against western temporal schemas that would see the 'post' as a stage in a linear epic in which the human overcomes itself and 'Nature' in order to become Anthropocenic, Colebrook sees the human as having always been posthuman from the very beginning, the subject of a 'Nature' that is permanently out of step with itself, always disturbing and terraforming its milieu. Colebrook's position poses rigorous questions about what is actually meant or desired when key terms in the posthumanist lexicon are referenced: entanglement, relationality, emergence, flat ontology, actants, composting etc. These are words that often come loaded, existing as place holders for a posthuman project of environmental repair that would seem so tangible, if only the leap beyond the fictitious and damaging exceptionalism of the human could be finally made. In contrast with these palliative performatives, Colebrook provides a sobering lesson, one that is not without suffering and cruelty. For her, the imperative is to avoid perpetuating what she brilliantly, if perversely, coins as 'posthumanism's ultra-humanism' (Colebrook, 2015: 225). Only then will it be possible to advance, tentatively, a progressive posthumanist ecology, one in which the human *is part of* the earth primarily because it is also positioned as being *apart from* it. Nothing – no God, idea or machine – will be able to suture the wound of this agonised 'a-partness', to impose an impossible totality. All the more so since the earth itself is constantly absconding from itself, always splitting, never whole – a kinetic process. In his *Theory of the Earth* (2021), the geophilosopher Thomas Nail notes:

> If the earth is a non-uniform and turbulent mover, as I argue, then the movement from A to B is much more like a continual transformation of the whole line AB itself. The earth is not uniform. Its movement is turbulent, unstable, and entangled with the cosmos in ways that we are only now discovering. This has radical and undertheorized consequences for our understanding of the earth and of motion. (Nail, 2021: 11)

In order to combat what Isabelle Stengers terms the destructive 'loneliness' of the human being' (Stengers, 2012), any viable theatre ecology today will have to be posthumanist one, albeit in a way that keeps Colebrook's complication firmly in mind. This is an ecology attuned to ghosts, remainders and 'parasites' – all of those enigmatic, resistant 'things' that stain the faith that Enlightenment humanism puts in the powers of rationality and reason to construct reality according to its own selectively imposed criteria. As Michel Serres remarks:

> The Cartesian meditation eliminates, expels, banishes everything hyperbolically. Once again, a clean slate and clear spot [...]. The thinking ego chases the parasites out [...] thus chases everything out, speaking absolutely; it discovers, elsewhere, the world, the white of our dominance. (Serres, 2007: 180)

As part of a western colonialist culture that has for so long sought to chase all 'parasites' out in the name of some dominant 'whiteness', the question that remains, for Serres, is how to diversify the world, to decolonise knowledge. Botching the entire operation is a real possibility, especially when one is cognisant of the fact that the 'parasite' is not a derogatory organism in Serres's lexicon, but an actant whose presence is necessary for the health of a planetary commons, the organism that shares the table with the 'host', that puts bodies into relation (Serres, 2007: 7).[10]

But as Colebrook has shown, not every posthumanist discourse is an ecological one, and, just as crucially, not every posthumanist ecology shares the politics, ethics and aesthetics that I am looking to affirm in my idea of a theatre ecology. In order, then, to define with greater precision how I understand posthumanism, it is useful to chart a brief course through the manifold approaches to the term that are often unhelpfully conflated in much ecocritical thought, and which many scholars and practitioners in Theatre and Performance Studies have yet to assess in any kind of critical depth, seduced as they sometimes have been by an impatient desire to import ideas from the outside without reflecting on what the larger consequences of these ideas are and/or their agendas entail (see Chapter 1, pp. 57–60).[11]

Posthumanist ecologies: monsters, bodies and politics

Quite appropriately given its critique of foundationalism, posthumanist thinking has multiple origins. Some of these are found in third-wave systems theory of Francisco Varela and Manuel DeLanda, some in poststructuralist philosophy, some in feminist science studies, some in new materialist philosophy, some in animal studies, and some in Indigenous thought.[12] Key thinkers include Donna Haraway (1991, 2003, 2016), Katherine Hayles (1999), Stacy Alaimo (2010, 2014, 2016), Jane Bennett (2010), Cary Wolfe (2010), Rosi Braidotti (2013), Jeffrey Jerome Cohen (2015) and Anna Lowenhaupt Tsing (2015, 2017).[13] In the contemporary posthuman landscape, Haraway's techno-aesthetic idea of the 'cyborg',[14] her 1980s allegory for border-crossing and entanglement, remains a central figure, especially amongst those posthumanists who point out the convergences between biological and computational life.[15] However, writing in 2016, and aware of a planet in crisis, Haraway is keen to 'earth' the cyborg, stressing its affiliations with the agentic play of matter and its proximity to creaturely life as opposed to its earlier technological affiliations:

> Cyborgs are kin whelped in the litter of post-World War II information technologies and globalized digital bodies, politics, and cultures of human and not-human sorts. Cyborgs are not machines in just any sense, nor are they machine-organism hybrids. In fact, they are not hybrids at all. They are, rather, imploded entities, dense material semiotic 'things' – articulated string figures of ontologically heterogeneous, historically situated, materially rich, virally proliferating relatings of particular sorts, not all the time everywhere, but here, there, and in-between, with consequences. (Haraway, 2016: 104)

In the ecologically fraught days of the Anthropocene, the temptation to continue with the cyborg as a hybrid apparatus for transcending flesh through technology, as one sees in the transhumanist fantasies of writers and theorists such as Vernor Vinge (1993) and Ray Kurzweil (2005), must be resisted.[16] To achieve that resistance, the cyborg is now figured as an interdependent creature whose identity is embodied – a way of thinking that is reflected in Haraway's turn to sympoiesis and composting as metaphors for understanding the microbial evolution of life:

> *Sympoiesis* is a simple word; it means 'making-with'. Nothing makes itself; nothing is really autopoetic or self-organizing. In the words of the Inupiat computer 'world game', earthlings are *never alone*. That is the radical implication of sympoiesis. *Sympoiesis* is a word proper to complex, dynamic, responsive, situated, historical systems. It is a word for worlding-with, in company. Sympoiesis enfolds autopoiesis and generatively unfurls and extends it. (Haraway, 2016: 58; original italics)

Theory 85

To bring the cyborg down to earth as a symbiont, something fleshy and biodegradable, is to return to 'ancient forms of monstrosity that modernity tried to extinguish' in its ecologically disastrous attempt to sever itself from everything deemed 'primitive', 'magical', 'animalistic' (Tsing, Bubandt, Gan and Swanson, 2017: 2).[17] This why the editors of *Arts of Living on a Damaged Planet: Monsters and Ghosts of the Anthropocene* (2017) are drawn to monstrous shapes and figures:[18]

> Monsters are useful figures with which to think the Anthropocene, this time of massive human transformations of multispecies life and their uneven effects. Monsters are the wonders of symbiosis and the threats of ecological disruption. Modern human activities have unleashed new and terrifying threats: from invasive predators such as jellyfish to virulent new pathogens to out-of-control chemical processes. Modern human activities have also exposed the crucial and ancient forms of monstrosity that modernity tried to extinguish. [...] Monsters ask us to consider the wonders and terrors of symbiotic entanglement in the Anthropocene. (Tsing, Bubandt, Gan and Swanson, 2017: M2)

Contesting the engrained anthropocentrism inherent in Francis Fukuyama's linear and dystopian understanding of posthumanism in his 2002 text *Our Posthuman Future: Consequences of the Biotechnology Revolution* (Fukuyama, 2002),[19] the movement back to the monster undoes a whole series of unhelpful binaries between human and non-human, organic and inorganic, mind and matter – something that materialist feminists have been particularly invested in for more than two decades now.[20] In her parsing of Elizabeth Wilson's *Gut Feminism* (2015), Vicki Kirby draws attention to a decidedly fleshy notion of philosophy, in which thinking is intestinal, a metabolic materialism that locates language and writing not in human language or computer code, but within the internal, cavernous workings of the body itself. To think with the gut, Kirby says, is to realise, quite literally, that the enteric system – eating, digesting, defecating – is 'always already social, psychological, cognitive, and mindful' (Kirby, 2017: 20), an inverted, shit-stained metaphysics that 'delivers a coup de grace to the humanist subject who sees herself as a pilot within the mere container of her body' (Kirby, 2017: 19).

Kirby's and Wilson's posthumanist return to the 'bowels of being' – what Myra J. Hird and Kathryn Yusoff, with an amusing nod to Guattari, call a 'line of shite' (Hird and Yusoff, 2019: 265) – thwarts western humanism's compulsion to purify itself of anything that would stick to or stain its metaphysics of whiteness: the earth, animals, peoples of colour, Indigenous communities, women who give birth and bleed, queer subjects that refuse to reproduce along normative lines.[21] In doing so, posthuman ecology offers a critical politics that historicises Leonardo da Vinci's 'unmarked' image of Vitruvian Man, a humanist representation that, as Rosi Braidotti points out, functions as 'an ideal of bodily perfection, which, in keeping with the classical dictum *mens sana in corpore sano*, doubles up as a set of mental discursive and spiritual values' (Braidotti, 2013: 13). Highlighting, like Serres, the close link that ties western humanism to imperial conquest, Braidotti explains how the so-called universalism of Vitruvian Man acts as an alibi for planetary destruction:

> Equal only to itself, Europe as universal consciousness transcends its specificity, or, rather, posits the power of transcendence as its distinctive characteristic and humanistic universalism as its particularity. This makes Eurocentrism into more than just a contingent matter or attitude; it is a structural element of our cultural practice, which is also embedded in both theory and institutional and pedagogical practices. [...] Central to this universalistic posture and its binary logic is the notion of 'difference' with pejoration. Subjectivity is equated with consciousness, universal rationality, and self-regulating ethical behaviour, whereas Otherness is defined as its negative and specular counterpart. (Braidotti, 2013: 15)

While Braidotti is certainly attuned to the violence of the western humanist project, its metaphysical and territorial desire to impose itself everywhere, she is nevertheless careful to insist that posthumanism is not an antihumanism. Posthumanism does not want merely to wash away the image of man from the shore of history as some mutant species (as Michel Foucault proposed in a celebrated image from his 1966 text *The Order of Things: An Archaeology of the Human Sciences*).[22] And neither does it seek to leave the earth behind, as one sees in the posthumanist fantasies of those ecomodernist thinkers who are either gearing up to terraform the earth or to leave it behind, completely, in some digital rapture, the final voyage of spaceship earth.[23] Rather, Braidotti's fleshed and feminist posthumanism is an ecology that commits itself to an ethics of corporeal difference, even as it recognises that the human is no longer the central agent on the earth. As Simone Bignall and Braidotti put it in the collection *Posthuman Ecologies: Complexity and Process after Deleuze*:

> The 'posthuman turn' – defined as the convergence of posthumanism with postanthropocentrism – is a complex and multidirectional discursive and material event. It encourages us to build on the generative power of the critiques of humanism developed by radical epistemologies that aim at a more inclusive practice of becoming-human. (Bignall and Braidotti, 2019: 1)

Ecologically as much as ontologically, one can understand why Braidotti's inclusive posthuman project refuses to give up on the human. For if a 'shared vulnerability [...] in the face of common threats, is refuted' (Braidotti, 2013: 50), why should one care for anything? Why not simply enjoy the negative freedom of being nothing, luxuriating in a gratuitous world that makes no ethical claims and where one is free to become whatever non-human object one desires? However, such a 'flat ontology' is ethically dubious and politically useless.[24] It says nothing about the environmental damage caused by those becoming. As Stacy Alaimo highlights in her critique of Ian Bogost's notion of alien phenomenology:

> We could as alien phenomenologists wonder what it would be like to be a plastic bag or plastic bottle cap. Or we could consider the network of chemistry, capitalist consumerism, inland water ways, ocean currents, and addiction to high fructose corn syrup that have created the Great Pacific Garbage Patch. (Alaimo, 2014: 19)

To think ecology with and after posthumanism is more than merely affirming our affinity with the non-human world, a joyful capacity to experiment and shapeshift, a becoming alien. It means paying attention to the more 'vertical' aspects of ecology, facing up to the mess, accepting the agency of a planet that is destabilised by the violence that certain human beings inflict on it.

So that there is no confusion pertaining to the version of posthumanism that I am concerned with in this book, I want to stress that my notion of ecological realism has little in common with the speculative realism of Quentin Meillassoux's (2008) and Ray Brassier's (2007), both of whom seek to purge western philosophy of its sentimental attachment to human consciousness as the exceptional faculty that would somehow guarantee our survival. According to Meillassoux and Brassier, western idealist philosophy is predicated on what they critique as a 'correlational understanding of existence', in which knowledge is divorced from any access to the world outside of itself, trapped in its own representation. In the correlationist thinking of a philosopher, like Immanuel Kant, for instance, the human being has no way of encountering the real ('the thing-in-itself'). The most that can be said about the world is that 'it might be there'. One never really knows for sure if the world exists because reality, for the correlationist, is an intellectual projection, an idealism, in which the object conforms to the mind. As a result of this epistemological barrier, philosophy encourages us to stop talking about the non-human world and to content itself with the concerns and machinations of consciousness. In the process, 'Nature' is forgotten.

In order to contest the humanism inherent in such correlationist fantasies, Meillassoux and Brassier show that thought is perfectly capable of gaining access to a reality beyond the human. For Meillassoux, this is evident in the capacity that humans have to speculate on the beginning of the universe, to encounter what he calls the 'arche fossil' (Meillassoux, 2008: 10).

Or alternately, in Brassier's case, to go in the opposite direction and to imagine the possibility of extinction. To go beyond humanism, in these instances, is not, then, to give up on thought; rather, it is to take thought to its logical conclusion, to see that the mind is not there simply for humans to reflect themselves in – a sort of intellectual prison house – but a faculty that positions us squarely within an earth that is beyond our control. While there is – at least potentially – an environmental politics to Meillassoux's and Brassier's 'realism', it is one that is devoid of affect, emotion and liveliness. For them, the most logical thing to do is to face up to the fact that humans are part of a larger non-human universe, in which thought may carry on without them. This leads both of them to engage in a process of melancholic renunciation, in which human existence is accepted as a mere stage in the earth's history – one that is neither necessary nor unique. The earth has already existed without humans and will do so again. Thought's role is to inform us of that unpalatable truth, to do away with the idea that the earth can be kept at bay, removed from the picture.

Yet for all its attack on the hubris of western humanism, there remains a stubborn and normative anthropocentrism to Meillassoux's and Brassier's projects. For while they are able to think a world beyond the human, they do not seem to be touched or moved by this world. At all times, they remain central, thinking beings that approach the earth as an object to conceptualise, a destiny to accept.[25] Neither emotions nor feelings have any place in their realism, and neither do they say anything about suffering in the here and now, the fact that some deaths are easier to support than others. Ultimately, their phallogocentric attachment to conceptual thought offers no ethical or pragmatic basis for developing a viable and transformative ecopolitics. As the postcolonial scholars Déborah Danowski and Eduardo Viveiros de Castro argue in their brilliant reading of Lars Von Trier's film *Melancholia* (2011), what Meillassoux and Brassier forget in their argument against correlationism is that humans do *feel* responsible for others in the face of disaster and extinction. In their analysis of the penultimate scene in the film, in which the melancholic protagonist Clare seeks to assuage the anxiety of her young nephew by building a tepee in the face of an imminent planetary collision, Danowski and Viveiros de Castro make it clear that posthuman ecology is not just about ending but ending *well*.[26] Where Meillassoux's and Brassier's realist ontologies are able to contemplate, quite dispassionately, a world without humans, Danowski and Viveiros de Castro tie ecological posthumanism to bodies that are ethical because exposed and vulnerable to damage. For them, care is a *pathos*, a distress that opens us to others. The imperative is not to live disabused, but to alleviate suffering, to believe in the earth, to affirm life as a good in itself:

> In the very last seconds before the encounter, already inside the 'magic cave' the simulacrum of an indigenous tepee that she builds with her nephew, we see the melancholia in her face give way to a crease of fear. A mere reflex contraction, perhaps, but precisely for that reason an (un)equivocal *sign of life*. It is just this moment of fear and shock that seems to differentiate *Melancholia* from the apocalypse announced by Brassier. (Danowski and Viveiros de Castro, 2017: 38)

Against Brassier's and Meillassoux's apocalyptic nihilism, the posthumanist ecology that interests me is predicated on the capacity of bodies to be touched emotionally and corporeally by forces from the outside. In such a world, ethics and politics escape their traditional dependence on laws, institutions and administrative processes; they are matters of health, things to live by. And ecology, by extension, also changes in that shift, referring now to the capacity of humans to enhance what the philosopher Baruch Spinoza terms *conatus* – the creative impulse that all living bodies have to persist in existence and to experiment with their lives. In this posthumanist ecology, 'Nature' is affective, a repository for becoming, something that creates passions, that insists on being affirmed for its own intrinsic qualities, and which is unable to be known by thought alone.

The importance that posthumanist ecologists such as Braidotti, Alaimo, Haraway and Kirby place on pre-verbal affect has much to offer theatre ecology, aesthetically and politically. By targeting bodies, they underscore the very thing that theatre is: namely, an affective affair, an experience of being with others in an actual milieu, of undergoing the shock of sensation. But there is a key point of distinction, too; something that prevents me from endorsing their ideas uncritically. For all the onus they place on corporeal and transcorporeal affect, what Alaimo generically and usefully terms 'the contact zone' (Alaimo, 2016), these thinkers all seem oddly confident in the capacity of language and philosophical thought to bear witness to the a-signifiable affects the body undergoes, to express its passion. In their writing, it is rare to find the body making an appearance as flesh that struggles to express its discomfort in signs, that de-articulates itself, even in the queered autobiographical texts of Mel Y. Chen that attempt 'to interweave biopolitical considerations of immunity into an account of the particular animacies and alienations of heavy metal poisoning, rendered in the first person' (Chen, 2011: 265).

To borrow from the feminist scholar Rebekah Sheldon, the goal in much posthumanist writing is still to 'demonstrate' knowledge, not to express sensation (Sheldon, 2015: 208). The result, Sheldon explains, is to deprive posthumanist ecology of a 'lexicon for apprehending' the liveliness of bodies, their sensuous excess, their refusal to signify, to stop making sense. Ultimately 'what is missing' in this writing 'is volatility as a quality of relations' (Sheldon, 2015: 209). At their core, humans are posthuman not only because they possess exposed bodies but because there is a mute, objectless force *within* their ontological make-up that makes any kind of self-coincidence or oneness impossible, a more-than-verbal 'too muchness' – what Sheldon, after Julia Kristeva, calls the pre-oedipal 'chora', the remnant or trace of womb music (Sheldon, 2015: 211–16). The real scandal of posthumanist philosophy, to return to Colebrook (see pp. 81–3), is that 'we' are haunted by a dissonance that always outwits the intelligence. At the very heart of the human being, in its most intimate relations with itself, there is a pre-individual excess – an 'accursed share' (Bataille, 1988) – that it is impossible to temper and/or quantify cognitively and linguistically. Confronted with this chaotic surplus, language retreats, defeated by swirls of colours, frequencies, movements, tastes and smells. No discursive concept will save us from this suffering, this *jouissance*. Nothing that would make existence reasonable, offering balm to our distress in the face of the earth. There is always a non-human pressure that thwarts rhetorical proclamations, a chaotic 'noise' whose consequences are uncertain and indeterminate, an anxiety that eludes capture in language's web.

Differently from Haraway who urges us 'to stay with the trouble by learning [how] to be fully present' (Haraway, 2016: 1), the ecology in theatre ecology is about learning how to be *more absent*, more out of step with one's self.[27] The 'trouble' here is not simply outside; it is inside, too, an ontological restlessness or irritability that nothing can resolve or soothe. The best one can do, perhaps, is to accept that a type of melancholia, not mourning, is the order of the day, an out-of-stepness that no amount of sympoietic affirmation will ever be able to resolve or even temper. To stay with the trouble is not to get beyond it, a mere stage in some dialectic of repair, as Haraway sometimes seems to suggest; it is to find oneself displaced, undone and disorientated by an earth that is inside and outside the human at the same time.

Without ever absconding from the urgent and pragmatic need for concrete forms of environmental action, the posthumanist ecology that interests me is one in which the real contaminant is found within the human, the consequence of an anonymous and terrestrial life force that circulates *in* and *through* us. A way of dealing with this boundless, disorientating energy is required. Denying or occluding it is not enough. For this force is part of the human's earthly inheritance, a 'gift' of matter that posits the 'soul' as something molecular and corporeal, a capacity for being affected from the outside, an immanent transcendence. As I will show in my reflections on Lyotard's *oikos* in Chapters 3 and 6 (see pp. 110–13 and pp. 183–9), the 'soul' is not a theological term for some proper self that will overcome death; it is a faculty for feeling, a source of torment as much as joy. To live ecologically is to face up to the doubleness of the terrestrial 'soul', to try to accept, as best one can, the earth as an erratic force that cares little for human desires, and which is determined to fulfil its own violent destiny in five billion years from now, in the wake of the heat death of the sun. On such a planet, the eye is not transcendent, and neither is language salvific. Consolation can be found only in the affirmation of terrestrial existence, living a million different lives in this one, a willingness to be de-created, again and again.

The poet Lisa Robertson offers evocative insights into the ecology I am concerned with, when she states that, in reading, she does not look for mastery or meaning, but instead 'the promiscuous feeling of being alive', a kind of 'rhythmic dispersal' that submits to 'a ruinous foundering' (Robertson, 2012: 13). In this dispersal of identity and language, an unspoken and unsolicited exuberance emerges that produces an attachment to life, an energised openness to things, a terrestrial passion. This is not a solace of absolution, a dramatic denouement that would put an end to the existential sorrow that besets the human in its doubled a-partness. Rather, it is a momentary reprieve, dependent on the acceptance that one's fate is to live on a stuttering and unstable Earth, without origin or *telos* – a way of thinking that tempers the positivity of much posthumanist ecology while also impacting on the field of ecocriticism as a whole. In particular, it marks a major departure from the primary value that many scholars and practitioners in that field have long accorded to two literary practices: hermeneutics as a critical method for exposing contradictions and ambiguities, and narrative as the privileged vehicle for 'crafting conditions for finite flourishing on *terra*, on earth' (Haraway, 2016: 10).

Troubling ecocriticism: beyond narrative

As a consequence of its traditional dependence on making sense, reading for secrets and symptoms, ecocriticism has been primarily concerned to interpret 'Nature', even when the focus is purportedly materialist. Serenella Iovino and Serpil Oppermann, for instance, preface their influential 2014 collection *Material Ecocriticism* by insisting on how materialist methods allow readers to trace a series of agentic connections, to map what matter *means*:

> Reading into the 'thick of things', material ecocriticism aims to explore not only the agentic properties of material forms, whether living or not, whether organic, natural or not, but also how those properties act in combination with other material forms and their properties and with discourses, evolutionary paths from metals to bacteria, from nuclear plants to information networks. (Iovino and Oppermann, 2014: 4)

Iovino and Oppermann are far from alone in their desire to '*read* into the thick of things'. In fact, most ecocritics, including ones explicitly concerned with affect such as Stephanie LeMenager (2014), Alexa Weik Von Mossner (2017) and Nicole Seymour (2018), remain wedded to this logic. Regardless of whether they look at novels, films and/or performances, the method is invariably the same: to show how the text represents and organises material artefacts for the intelligence to decipher, as opposed to expressing how the haecceities of surface, rhythm and gesture exceed the sign. Ecocriticism's logocentric allegiance to interpretation is paralleled by an equally enthusiastic faith in the capacity of narrative to produce new discursive arrangements of the world. LeMenager concludes her book on *Living Oil: Petroleum Culture in the American Century* with the following panegyric:

> I end this book with the assertion that academic humanists in the United States and elsewhere, particularly those interested in the production of narratives across a variety of media, have something to contribute to a future that challenges Tough Oil. That 'thing' is narrative itself. We can disseminate it, and we can make it. (LeMenager, 2014: 184)

LeMenager's words highlight the quasi-messianic status of narrative in the Environmental Humanities. For further proof, consider this comment from Haraway who proposes a turn to Indigenous 'world games' as a way of remaining human in the face of environmental catastrophe:

> World games are made with and from indigenous peoples' stories and practices [...]. These games both remember and create worlds in dangerous times; they are worlding practices. Indigenous peoples around the earth have a particular angle on the coming extinctions and exterminations of the Anthropocene and Capitalocene. The idea that disaster will come is not new; disaster, indeed genocide and devastated home place, has already come, decades and decades ago, and it has not stopped. The resurgence of peoples and places is nurtured with ragged vitality in the teeth of such loss, mourning, memory, resilience, reinvention of what it means to be native, refusal to deny irreversible destruction and refusal to disengage from living and dying well in presents and futures. (Haraway, 2016: 88)

Or this from Anna Lowenhaupt Tsing, who speaks of the need for stories of 'contaminated diversity' that would contest capital's 'algorithms of self-containment':

> If a rush of troubled stories is the best way to tell about contaminated diversity, then it's time to make that rush part of our knowledge practices. Perhaps like the war survivors themselves, we need to tell and tell until all our stories of death and near-death and gratuitous life are standing with us to face the challenges of the present. It is in listening to that cacophony

of troubled stories that we might encounter our best hopes for precarious survival. (Tsing, 2015: 34)

Or this from Thom van Dooren, writing from what he terms the 'edge of extinction':

> [A]t the same time as they may offer an account of existing relationships, stories can also connect us to others in new ways. Stories are always more than simply descriptive: we live by stories and so they are inevitably powerful contributors to the shaping of a shared world. [...] I see storytelling as a dynamic act of storying the world, utterly inseparable from lived experience and a vital contributor to the emergence of what is. Stories arise from the world, and they are at home in the world. [...] As a result, telling stories has consequences: one of which is that we will inevitably be drawn into new connections, and with them new accountabilities and obligations. (van Dooren, 2014: 10)

As with Frank Kermode's classic 1966 study of eschatology *The Sense of an Ending: Studies in the Theory of Fiction*, the environmental stories proposed here are supposedly able to transcend their fictional status by 'mak[ing] sense of our lives' through the 'shap[ing] and organ[ising] of experience (Kermode, 2000: 4, 3).[28] They do so by 'composting' new temporal relations between past, present and future (Haraway, 2016: 136); highlighting places and moments where animals and humans meet; and offering insight and access into the messy and ambivalent complexities of life. 'Narrative', van Dooren confirms, 'is my way into this complexity; stories allow us to hold open simultaneously a range of point of views, interpretations, temporalities and possibilities' (van Dooren, 2014: 8).

Van Dooren's attachment to story is reflective of a larger trend in the Environmental Humanities. One that would stress the unique contribution literature can make to the climate crisis by supplementing scientific data with emotions and feelings – what Scott and Paul Slovic see as the necessary alignment of 'nerves and numbers' (Scott and Slovic, 2015). The literary scholar Alexa Von Mossner illustrates that alignment when she uses the environmentally evocative example of 'slipping on ice' to show how 'a narrative is different from a scientific account in that it will allow those who receive it to imagine a sense of what [that slipping] is like' (Von Mossner, 2017: 17). In this simulation of feeling, readers, Von Mossner contends, create empathetic bonds and affective pathways with people, events and things 'that continue to impact our emotions, attitudes and behaviours long after we have finished engaging with [them]' (Von Mossner, 2017: 7).

From a strategic perspective, one can understand why hard-pressed humanities scholars would want to invest in Von Mossner's practice of 'econarratology' (Von Mossner, 2017: 12). It carves out a specific role for the place of the imagination in a research field – ecology and environment – that has been historically dominated by 'hard' scientific discourses. While I do not dispute, in any way, the role of aesthetics in the construction of a more ecologically progressive world – this is the logic of my own argument, after all – I do nevertheless want to take issue with two unspoken assumptions in Von Mossner's argument. First, that narrative should be accorded such a primary role in provoking environmental thinking; and, second that environmental thinking itself is based on imaginative empathy, placing oneself in the shoes of the other, as it were.

As I argue throughout this book, neither *logos* nor story has a monopoly in making us believe in the earth. On the contrary, it may even be that narrative, in the way it invariably privileges the active powers of the human imagination to shape the world, could blind us to what really *matters*: the unspeakable haecceity of terrestrial things, an im/material force that remains elusive, hidden, non-verbal, something that needs to be suffered in and by the flesh. Here nerves are not synthesised with numbers; they are indicators of existential distress, a primal wound of sorts.[29] Narrative can often obscure this inarticulate power of things, because its aim, at least in the accounts I have discussed above, is to guide the reader through a coherent landscape of signs, to tell a story that 'will make sense', to elicit a telling phrase. In doing so, narrative often humanises the world, making it into an imaginary object, no matter how deconstructive it may sometimes be of the artificial cosmos it has created. As Kermode insists, 'the imagination is form-giving', 'a maker of order and concords', a technology that necessarily organises the chaos that comes before perception, the very shock that provokes thinking (Kermode, 2000: 144).

Thankfully, however, there are other ways of ecologising through aesthetics, alternate modes of representation. It is not just the empathetic imagination that counts in the creation of affective ties with the earth. Affective dislocation is also required, a de-phasing of the actual, the creation of a gap between feeling and cognition. Before we imagine, in other words, we need to be confronted with power that de-creates the world as it is, that sets it reeling. This power exists beyond the signifier and resists representation. It is a process not a thing.[30] The avant-garde film-maker Jean Epstein names this force 'the recoil before the leap, the moment before landing, the taut spring, the becoming, the hesitation, the prelude, and even more than all these, the piano being tuned before the overture' (Epstein, 1988: 236). Epstein's description allows one to see that imaginative thinking – like all thinking – is terrestrial. It does not only come into being through the presence of narrative structures that create textual experiences (fictional or otherwise) for a reader to recognise, identify with and then appropriate for their own ends; rather, it is based on pre-verbal affects – anonymous forces – that impress themselves on consciousness, while remaining opaque, beyond the reach of narration, ex-orbitant and unowned.[31]

For a theatrical ecology

While a concern with the opacity of affect is not absent in text-based, eco-critical theory – one thinks of the work of Sarah Wood (2014), Timothy Clark (2015), Claire Colebrook (2014) and Rebecca Sheldon (2015) – it seems particularly important for theatre and performance scholars to engage with. Theatre's sign systems do not appeal to the mind – at least not initially – as something to interpret, and nor do they present themselves in an ordered, discrete sequence as language generally does on a page. Theatre is an impatient medium in which 'everything happens all at once', creating a sensate whorl in which verbal language forms a mere part of a dense and dynamic vectoring of experience. In theatre, speech is always excessive, ingrained in bodies that materialise it by making it 'sound', provoking a kind of elusive 'noise-music' reverberating through time and space like the screams and glossolalias in Artaud's *To Have Done with the Judgement of God* (1947), the cries and chants of Marguerite Duras's actors in *India Song* (1973) and the guttural contortions and distortions in Diamanda Galás's voice as she sings *O Death* (2008), on a piano bathed in smoke and fire, a landscape filled with some strange, cosmic yearning.

As a way of remaining faithful to theatre's 'mattering', the theatre ecologist is tasked with *feeling* the ecological pulse that theatre always already possesses, to undergo what I call the 'test of theatricality' – a willingness to be double, to inhabit excess, to affirm the virtual. In my understanding, and I will explain this in greater detail in Chapters 4 and 5, theatricality is neither associated with creation of an artificial world, and nor is it equated with the foregrounding of appearance, the 'laying bare of the mediating process of reflexivity that gives rise to thought, text, and image', as Timothy Murray puts it in the landmark collection *Mimesis, Masochism, and Mime: The Politics of Theatricality in Contemporary French Thought* (Murray, 1997: 2–3). Rather, theatricality, as I use it, interrupts space, time and identity, creating an anonymous flux in which objects lose their names and forms and start to vibrate with a kind of molecular power, an im/material energy that leaves an 'impress' of its virtual passage on the bodies of those who attend to it. As Hans-Thies Lehmann puts it in his description of the sensory overload caused by *Tragedia Endogondia*, the cycle of performances staged across Europe by Socìetas Raffaello Sanzio in the early 2000s:

> [I]t is impossible to do justice to the wealth of auditory and visual inventiveness, scenes and images. It is a theatre of affectivity, not romantic emotion – one that is 'inhuman' and practically never shows the human being in dialectical exchange. (Lehmann, 2016: 433)

Here theatricality – Lehmann shows us – suspends the humanist need for direction and shape. The compulsion to follow a preordained narrative, as one finds in much dramatic theatre, is replaced in these inhuman or 'theatrical theatres' with a new attachment to the weight, density and volumes of objects and things. Through this attentiveness to haecceity and movement, spectatorial bodies become receptive, energised, ex-posed. To transpose the insights of Christopher Neve from landscape painting to performance, theatricality 'shows life not as a [narratological] development but as a condition' (Neve, 2021: 10; my addition). Its task is to materialise, to bring us to our senses, in the most literal of ways by creating jerky syncopations in which the im/material force that all objects harbour within themselves is released outwards like an extruding wave. To follow that ripple, dramatic tension needs to retreat and narrative arcs shatter. Everything has to be attuned to the plasticity of surfaces, to what they show and hide, to how they affect. I think of Gisèle Vienne's slow-motion dancers in *Crowd* (2017), moving the air and drawing the audience into their orbit, sticking to them; Samuel Beckett's actors pacing the stage in *Quad* (1981) and *Footfalls* (1975) and communicating through rhythm; Goat Island's performers standing still and trembling in *When Will the September Roses Bloom – Last Night Was Only a Comedy* (2005), making us think by remaining so stubbornly there, so apart and alone.

Theatricality draws from and runs on what Jean-François Lyotard terms the '*gestus*' (Lyotard, 1993b: 38; original italics). As I explain in Chapter 6, the *gestus* is not synonymous with what is commonly understood as a gesture. Rather, it is a virtual force – a gest or jet – that resides within the dense and opaque materiality of bodies, an enigma or dark shadow that Lyotard associates with the 'pressure of things', and that nothing can 'prepare' us for (Lyotard, 2009a: 19). Anticipating Artaud's ideas on mise-en-scène that I discuss in Chapter 5, the goal of this gestic theatre is not to produce representations that would rehearse the future as an object to identify with or hopeful horizon to move toward.[32] The logic is more basic: to release the *délire* that is already at work in matter, to create becomings that no imagination can temper or domesticate, to get to the creative dissonance in bodies that provokes new enthusiasms for life in an extra-linguistic spacetime. To leave a performance by the Butoh artist Kazuo Ohno is to exit transformed, to have experienced multiplicity in action, to feel oneself haunted by the ghosts of others that silently inhabit us, that keep us company, that tempt us to become other, that take us beyond the human.

The vitalist philosopher Gilbert Simondon terms this generative exuberance the 'pre-individual'. For him, the pre-individual is the source of all individuations and becomings in life, a desire for form that already '*preexists in the system*' but which the system can neither capture nor contain (Simondon, 2020: 6, 13; original italics).³³ 'Pre-individual being', Simondon insists, '*is being that is more than unity and more than identity*', an impulse that infuses the actual with the virtual, with the possibility of difference from the very beginning, a relentless perturbation that exists *within*, and not just *between* objects (Simondon, 2020: 5; original italics). As hosts for the pre-individual, individuated bodies and organisms are temporalised, performers on a terrestrial stage that they are unable to exit from, caught up, as they are, in a movement of becoming where boundaries, positions and functions are blurred: '*The living being is a system of individuation, an individuating system, and a system that is in the midst of undergoing the process of individuating*' (Simondon, 2020: 7; original italics).

For Simondon, the crucial thing, the most profound question, is not to discover what things are but to track the molecular, indeterminate process by which they become. To do that, Simondon replaces induction and deduction with 'transduction', a term that denotes a 'physical, biological, mental or social operation' – in which 'an activity propagates incrementally' as it moves from one 'structuration to another', all the while using that structure or domain as the 'principle and model' for the next iteration (Simondon, 2020: 13). In this move from static being to dynamic becoming, individuation is characterised by 'phasings and de-phasings' (Simondon, 2020: 3, 4). These de-phasings are spatio-temporal bifurcations that allow the organism to transform itself from within, to give itself a sense of duration in which nothing is lost or sublated. Importantly, Simondon claims that one can experience a de-phasing only by participating in it, which, of course, means that one's perspective is invariably partial, incomplete and processual:

> The individuated being is neither the whole being nor the first being; *instead of grasping individuation on the basis of the individuated being, the individuated being must be grasped on the basis of individuation and individuation on the basis of the pre-individual being*, which is distributed according to several orders of magnitude. (Simondon, 2020: 12; original italics)

In this ontogenetic process, this intensive gathering of multiplicities and scales, ecology *becomes* theatrical, a concrete, spatio-temporal distribution in which individual organisms become other than what they are by opening themselves to a pre-individual excess – a surplus, an internal itch – that is neither quite inside nor outside, but intermediate, a disjunctive conjunction.

No rehearsal or script can contain the unpredictability of this excessive force, 'this dark zone' that catalyses phase-shifts (Simondon, 2020: 10). Across all its de-phasings, the pre-individual has no fixed terminus to achieve or plan to fulfil. Its sole purpose is to actualise itself, to find a body in which to individuate and mutate, to become im/material. The pre-individual's appetite is insatiable. Its only desire – and for Simondon, this is a theatrical desire – is to experiment with life. It is content to begin again and again, to squander its resources in the knowledge that there is always more to come, another performance to produce, a different individuation to accomplish. Elizabeth Grosz explain why:

> The preindividual is both the precondition of any individuation, and thus of any individual, but also the extra 'charge' that individuation carries within it as it develops and elaborates new orders to address new kinds of problems, a resource for ongoing individuations that may occur within and between individuals. The preindividual may be understood as the indistinguishably mental/material condition for thought and things, mind and matter. It makes every individual, material or mental, living or non-living, possible. The preindividual is neither material nor ideal but the dynamic force, the charge of potential, that enable both to come into being and to function in increasing interrelations and orders of complexity. (Grosz, 2017: 174)

In the same way that any theatre production irreversibly changes its source text or score in the very act of staging itself, so concrete individuations, as Grosz intimates, do not simply follow a linear movement from cause to effect. As well as being informed by their context or milieu, they remain in contact with the pre-individual energy they originated from, finding new potential within it, allowing it to return differently. Underscoring the necessity for the parallel I am drawing between theatre and ecology, Simondon stresses, quite blatantly, that 'the living being conserves within itself an ongoing activity of individuation; it is not merely the result of individuation, like the crystal or molecule, but a *theatre of individuation*' (Simondon, 2020: 7; my italics).[34]

In Simondon's theatre of individuation, life is seen to be out of step with itself, an operation or drive whose actualisation is only a partial manifestation of some greater potential, an endlessness that is indeterminate and infinitely plastic, a de-phasing movement that resists totality and closure. On Simondon's stage, individual entities are metastable, incompatible with and to themselves. Their 'internal resonance', the vibration of the pre-individual within them, means that they are unable to identify, fully, with the roles they have been accorded. There is always something else to come, a 'wealth of potentials', resulting from the lack of a substantiated identity or essence to their being. Simondon notes:

Furthermore, unlike that of the physical individual, the whole activity of the living being is not concentrated at its limit; in the living being there is a more complete regime of *internal resonance* that requires ongoing communication and that maintains a metastability, which is a condition of life. This is not the only characteristic of the living being, and the living being cannot be compared to an automaton that would maintain a certain number of equilibria or would seek compatibilities [...]; the living being is also a being that results from an individuation and amplifies this individuation, which is something that is not done by the technical object to which cybernetic mechanism would want to functionally compare *it*. (Simondon, 2020: 7; original italics)

Simondon's ideas effect a stunning reversal in how theatre's relationship with ecology has been conventionally approached. Theatre is ecological not because it represents environmental ideas, but because ecology itself – Simondon has already told us – is theatrical, a process that is defined by diffractions, incompleteness and transformations. In this performative configuration, the earth is imagined as one vast, seething theatre, a planet crisscrossed with elemental energies and in constant movement. Critically, this earthly drama is not an updated version of the *theatrum mundi* idea that was so prevalent in early modernity, in the work of Shakespeare and Calderón de la Barca. Where the *theatrum mundi* is ruled by a director who stands apart from 'his' (*sic*) creations, watching them play out their prescribed destinies from afar, theatre ecology is part of a more intimate and immanent performance. It belongs to what Georges Bataille calls a 'general economy', an earth that is open to invisible energies of the cosmos, to the chaotic economy of the sun (Bataille, 1988: 19–28).[35] In Simondon's theatre of the pre-individual, everyone and everything acts on each other. There is no absolute outside either to theatre or ecology, and neither is the relationship reflective. All is fractal and processual – ecology is theatricalised and theatre is ecologised. No superlunary God or director organises this theatre ecology; and no human can stand apart from it and control it. Like theatre, ecology is a virtual force, a property of the pre-individual's theatrical drive to experiment with bodies that it depends on, but which it is never identical with and always in excess of. And like ecology, theatre exists to affirm the deterritorialisations of the earth, to advocate for the contingency and plasticity of flesh, the drive of life to move beyond oneself. Taken together as a composite term, theatre ecology wagers on the posthuman. It affirms the cruel realisation that human existence is neither eternal nor necessary, but as gratuitous and momentary as the performance of a play or the evolution of an organism, that is perpetually coming into being.

To approach theatre as an ecological art of incompleteness, a staging of life's disparity, means that there is no need for it to represent environmental issues in narrative form, or to confect a sentimental idea of 'Nature' to 'fall in love with', as actor-director Mark Rylance put it, in an article in the *Guardian* newspaper in June 2021. All of this is too anthropocentric, too wilfully focused on making connections with a 'Nature' that was never there, in the first place. To avoid this 'pathetic fallacy', environmentally concerned theatre practitioners ought to experiment with the construction of materialist scenographies and dramaturgies that use the affective power of bodies to create affects and sensations that make individuals indiscernible, that unsettle the ground on which their identity is built.[36]

As Simondon reminds us, 'affectivo-emotivity is not merely the reverberation of the results of action within individual being; it is a transformation; it plays an active role' (Simondon, 2020: 279). Affect is transformative because it belongs to the pre-individual, and precisely because it belongs to the pre-individual it must remain anonymous, always less and more than what it appears to be. Affect's anonymity propels individuals outside of themselves into collective worlds. It forms the basis of new thoughts, values and actions. It glues us to life; creating responsibilities and appetites for becomings: 'Affectivity', Simondon maintains, 'is what leads the charge of pre-individual nature, the mediation between that which is pre-individual and that which is individual; it is the manifestation and reverberation in the subject of the encounter and emotion of presence, of action' (Simondon, 2020: 279). If Simondon's ideas are accepted, then it does not matter what theatre says or shows about the environment representationally. The heart of the matter is elsewhere; in how practitioners are able to use the medium to release the affective power of the pre-individual, a terrestrial force that forms attachments to the earth through de-phasings, not discourses. The aim, in other words, is to expand the aesthetic, approaching ecology as a theatrical phenomenon in and by itself, a creative impulse that produces life by dislodging fixed notions of space, time and identity in such a way that a becoming is generated.

This all-too brief parsing of Gilbert Simondon's notion of 'a theatre of individuation' marks an important stage in the theorisation of a specifically *theatrical* model of theatre ecology, one that is rooted in metonymy, not metaphor. Approached as a part of a larger, theatricalised process, a present-ness without a present, theatre has no need to stage ecology because it *is* ecology, a corporeal medium implicated, empirically, in the always already performative transformations and elaborations of pre-individual life. To participate in that theatricalisation is where theatre's contribution to the creation of a more life-affirming ecology resides.

Conclusion

In this chapter, I have sought to specify what the word ecology in theatre ecology actually refers to. While my ecology is certainly a posthumanist ecology, it diverges theoretically and aesthetically from numerous contemporary models of thinking the posthuman. Theoretically (or ontologically), I have been concerned to emphasise the affective 'distress' of ecology, the sense in which ecology is about coming to terms with a strange, non-human dissonance within subjectivity that nothing can salvage, repair, or represent; and aesthetically I have argued for the need to move away from ecocriticism's over-investment in interpretation and narrative in order to grasp, instead, the theatrical quality of ecology, the sense, in which like the theatrical medium itself, it deals with the affective power generated by bodies in the process of becoming other than what they are. To remain faithful to that power, it is important to pay attention to intensities and sensations, to perceptions that resist meaning but which demand a response – one that is able to think alongside and co-create with them. Gilbert Simondon's notion of a 'theatre of individuation' played a key role in this shift, since it posits ecology as an always already theatrical process, a terrestrial performance in which pre-individual forces need individual organisms to experiment with and de-phase from. Simondon's ideas are only a first step in this study.

What remains to be created in the chapters to come is a theory attuned to the specificity of theatre ecology. One that would be alive to its aesthetics and politics. For, as I argued in the Introduction, theatre ecology cannot just be content to posit a 'new earth', it calls out, too, for a planetary people to inhabit it, a bastard, creolised 'commons' that is willing to oppose the capitalist desire for territory and to wager instead on terrestrial becomings. To underline that point, I finish this chapter with some pertinent words from Glissant:

> Ecology going above and beyond its concerns with what we call the environment, seems to us to represent mankind's drive to extend to the planet Earth the former sacred thought of Territory. Thus, it has a doubled orientation: either it can be conceived of as a by-product of this sacred and in this case it will be experienced as mysticism, or else this extending thought will bear the germ of criticism of territorial thought (of its sacredness and exclusiveness), so that ecology will then act as politics. (Glissant, 1997: 146)

Notes

1 According to Haeckel, who invented the term in 1886, ecology is the study of the 'economies' of living forms, a discipline dedicated to measuring how animal

organisms metabolise energy. Particular emphasis in Haeckel's definition is placed on the relationship between organism and environment, the way in which living things are invariably part of and adapted to a larger system (*Umwelt*).
2. A keyword is a word with disparate and often dissensual meanings.
3. For an excellent, in-depth genealogy of the early first wave of environmental philosophy and ecocriticism, see George Sessions (1987).
4. I provide a concise definition of autopoesis on p. 72, note 28.
5. Queer intimacy undoes hierarchical notions of intimacy that make ontological and ethical distinctions between different types of substances. In queer intimacy, one can be intimate with all sorts of earthly materials: other humans, animals, rocks, textiles, metals.
6. In a footnote in *The Environment on Stage: Scenery or Shapeshifter?* Julie Hudson speaks of four waves of ecocriticism as a way of accounting for the disparate orientations of contemporary theory (Hudson, 2020: 10).
7. Chakrabarty has been criticised for upholding a universalist notion of 'species being' in which everyone is considered the same, even if that 'universal', as Chakrabarty puts it, is 'a negative universal' (Chakrabarty, 2009: 222). The flaw here is that species being is still thought of as something fixed and continuous, a totality opposed to dynamics of historical being. A more progressive way of contesting the overarching sameness of life that Chakrabarty assumes is to see 'species being' as impossible being. Species being is not continuous, it is metastable, never quite itself. It is not a unified identity that we all share.
8. Nancy's term comes from his attempt to make sense of the Fukushima disaster in Japan in 2011, in which a meltdown of the nuclear power plant was caused by an earthquake that trigged a tsunami.
9. In numerous texts, but most importantly in *We Have Never Been Modern* (Latour, 1993), *Politics of Nature: How to Bring the Sciences into Democracy* (Latour, 2004) and *Facing Gaia: Eight Lectures on the New Climatic Regime* (Latour, 2017), Latour contends that modernity is defined by a desire to separate humanity from 'Nature', to create a self-sufficient notion of the human.
10. As Serres put it in *The Parasite*, 'There is no system without parasites. This constant is a law' (Serres, 2007: 12).
11. A notable exception is Christel Stalpaert *et al.*'s edited collection *Performance and Posthumanism: Staging Prototypes of Composite Bodies* (Stalpaert, Van Baarle and Karreman, 2021), which was published as I was editing this book.
12. For four excellent publications that trace the diverse origins trajectories of posthumanism, see Carey Wolfe (2010), Rosi Braidotti (2013), Stefan Herbrechter (2013) and Pramod K. Nayar (2014).
13. I list only some of the texts of these important and prolific theorists.
14. In her 1985 text 'A Manifesto for Cyborgs: Science, Technology and Socialist Feminism in the 1980s', Haraway stressed the technological aspects of the cyborg as she was keen to disrupt patriarchal discourses that associated women with 'Nature' in order to essentialise and repress them. The cyborg has no essence; she is constructed, a human machine. For more on this point, see Haraway (1991).

15 See, for instance, Katherine Hayles's proclamation that 'there is a striking correspondence between the conditions under which [biological life] is likely to emerge and those under which computation is likely to emerge – a convergence regarded by many researchers as an unmistakeable sign that computation and life are linked at a deep level. In this view, humans are programs that run on the cosmic computer' (Hayles, 1999: 241). Despite these comments, Hayles's posthumanism is suspicious of the celebratory claims of technophile, cyborgian thinkers who would see the computer as a way of leaving the human behind. To reject the body, for Hayles, is to reject politics and the earth, to believe in a spurious and problematic transcendence that denies the very source of how humans come to make sense of the world, and indeed to care for it. Hayles finishes her book in the following way: 'Although some current versions of the posthuman point toward the antihuman and apocalyptic, we can craft others that will be conducive to the long-range survival of humans and of other-life forms, biological and artificial, with whom we share the planet and ourselves' (Hayles, 1999: 291).

16 Déborah Danowski and Eduardo Viveiros de Castro equate transhumanism with the Singularity Thesis advocated by Vinge in his essay 'The Coming Technological Singularity: How to Survive in the Posthuman Era' (Vinge, 1993). Singularity depicts a world without bodies or place, a world of superhuman intelligence, in which humans would interface with computers. The logical conclusion of Singularity is that death is abandoned, since what matters is no longer the existence or health of the body but the data generated by the mind alone. That data can be stored on servers, available to be downloaded in the future. In Singularity, immortality beckons. We no longer need the earth, a supercomputer is all that is required. This is how Danowski and Viveiros de Castro put it: 'the culmination of the Anthropocene brings with it the obsolescence of the human […].We will no longer be accountable to the world, we will no longer have to deal with any limits, because we will have become the world by turning the world itself – the cosmos as a whole – into a 'magnificently sublime form of intelligence' (Danowski and Viveiros de Castro, 2017: 47).

17 For more on the 'ecology' of the monster, see Jeffrey Jerome Cohen's early intervention *Monster Theory: Reading Culture* (Cohen, 1996) and Bruno Latour's figure of the 'Cosmocolosse' in *Facing Gaia: Eight Lectures on the New Climatic Regime* (Latour, 2017: 9–14).

18 These inclusions go by several names: 'planetarity' (Spivak, 2014), 'indigenous conceptual worlds' (Viveiros de Castro, 2016), 'sympoiesis' (Haraway, 2016: 97–8), 'Gaia' (Latour, 2017) and 'pluriverses' (Escobar, 2018).

19 Fukuyama makes the mistake of regarding the posthuman as something that comes after the human, not as a possibility that the human already harboured with itself, a post that came before.

20 Fukuyama's aim is to defend the human from the encroachment of artificial life. Unlike Colebrook, what he fails to see in the linear line he traces from humanism to posthumanism is that the human has always been posthuman from very beginning. Its outside is already inside.

21 For more on impurity, bodily waste and posthuman ecology, see Haraway on composting (Haraway, 2016: 134–68) and Jennifer Mae Hamilton and Astrida Neimanis's article 'Composting Feminisms and Environmental Humanities' (Hamilton and Neimanis, 2018).
22 This focus on the posthuman rather than the anti-human marks an important divergence from Verena Andermatt Conley's important early theoretical intervention into what a posthumanist ecology could be in *Ecopolitics: The Environment in Poststructuralist Thought* (Andermatt Conley, 1996: 40–55). In that text, structuralism, in particular Claude Lévi-Strauss's anthropological version, is the touchstone for a new ecological ethics, precisely because it promises to do away with 'man'.
23 In the digital rapture, technological systems and human biology 'will come into fusion with one another, generating a superior form of machinic consciousness that will nonetheless be at the service of the human will' (Danowski and Viveiros de Castro, 2017: 46).
24 Flat ontology describes a mode of being in which everything is of equal weight and importance. As such, flat ontology is unable to predicate value or to distinguish between who is guilty and who is innocent. While the flat ontologist can be amazed by the strangeness of things, as Ian Bogost is in his text *Alien Phenomenology, or What It's Like to Be a Thing* (Bogost, 2012), they cannot account for the suffering imposed by certain bodies on others. Flat onology lacks an ethics, politics and history.
25 The idea of thought as something separate from humans highlights a parallel between Meillassoux and Brassier and technomodernists such as Vinge and Kurzweil.
26 Danowski and Viveiros de Castro point out that the cosmopolitical motto of Amerindian peoples, like Derrida's ghost, is '*vivir bien, no mejor*' (to live well, not better) (Danowski and Viveiros de Castro, 2017: 76).
27 If one accepts that the human is always haunted by the non-human, internally and externally, then it is difficult to know how 'one could ever be fully present', as Haraway asks us to be. In fact, to come into contact with the trouble of the non-human is to accept a sense of exile from one's own humanness.
28 A very different, but no less curative sense of ending is apparent in the cosmologies of Amerindian cultures. In Amerindian narratives, not only is space left for numerous endings, but these endings of the world are never without people, in contradistinction to the western eschatological tradition. As Danowski and Viveiros de Castro note: 'Indigenous people have something to teach us when it comes to apocalypses, losses of world, demographic catastrophes, and ends of History, for the native peoples of the Americas, the end of the world has already happened – five centuries ago. To be exact, it began on October 12, 1492' (Danowski and Viveiros de Castro, 2017: 104).
29 It is worth pointing out that 'nerves' by themselves are not necessarily progressive, especially when it comes to thinking through what a democratic political ecology might consist of. As the Tiqqun Collective point out, nerves can also be used to maintain a permanent state of collective anxiety and crisis in the social

body, thus making it docile and respons-able in a decidedly reactionary manner to what they call '*Imperial governance*' (Tiquun, 2010: 9; original italics). This cautionary insight needs to be applied to all affective ecologies.

30 This does not mean that language is devoid of affect; it is not, but its capacity to tap this anonymous energy is far from self-evident. Language and narrative often have to fall apart to express these terrestrial powers, to work against themselves, to un-say. As Lyotard notes: 'Words "say", sound, touch, always "before" thought. And they always "say" something other than what thought signifies. [...] Words want nothing. They are the "un-will", the "non-sense" of thought, its mass' (Lyotard, 1991: 142).

31 For more on the non-verbal power of theatre to create ecological affects, see Lavery (2013, 2015, 2016b and 2019a).

32 Artaud wants mise-en-scène to avoid mere textual representation and instead to exist as a 'poetry of space independent of spoken language'. The intention, always, is to create 'intensity', sensations that elude figuration (Artaud, 1976: 233).

33 For an excellent essay that looks at the intimate relationship between Simondon's notion of individuation and theatre, see Sally Jane Norman (2012). Although Norman is aware that 'theatre might thus be considered as a kind of technical-cultural apparatus whose role is to maintain the multiple, polyvalent, potential trajectories that can quicken the emergence of imaginative new paradigms', she does not attach that becoming or emergence to ecology (Norman, 2012: 126). Someone who does is Erin Manning in her writings on Simondon and dance (Manning, 2013: 17–20).

34 Where the crystal comes to an end point once it has exhausted the pre-individual potential within its 'supersaturated mother liquor' (Simondon, 2020: 13), the living being shows itself to be 'both superior and inferior to unity', 'doubled by an on-going individuation' (Simondon, 2020: 9). To play on Simondon's metaphor, the living being is always on stage.

35 In *The Accursed Share, Vol. 1: Consumption*, Bataille argues that the restricted economy of Marxism is part of a larger, more general or ecological economy whose source is the sun. This is an excessive economy that, as with Marcel Mauss's notion of the gift, is based on expenditure rather than reinvestment (Bataille, 1988: 63). Surplus value – the accused share – always needs to be squandered. For an important ecological reading of Bataille, see Allan Stoekl's *Bataille's Peak: Energy, Religion, and Postsustainability* (Stoekl, 2007).

36 The pathic is contrasted with the empathic in this relationship. Where empathy looks to identify with the loved object, to construct a unity, the pathic confronts the subject with things that it is unable to identify with. In the pathic, the object retains its otherness and difference.

3

Model (a lexicon for theatre ecology)

Introduction

Simondon's ideas on individuation highlight the need to theatricalise ecology, to see life itself as a stage for becomings. Nevertheless, for all the onus he places on performativity, Simondon says nothing about how theatre, as a specifically aesthetic medium, might be able to feed into that deeper and vaster process. And neither does he say anything substantial about politics – the very thing that decolonial ecologists such as Édouard Glissant and Malcom Ferdinand insist on. In order then to respect the expansiveness of theatre ecology, I construct a theoretical model in this chapter that borrows from the work of Karen Barad, Jean-François Lyotard and Félix Guattari – three posthumanist philosophers who, in their different ways, are cognisant of the need to approach ecology as an aesthetics and politics that is both human and non-human at the same time. More direct in their critiques than Simondon, they are sensitive to how capitalist modernity's military-industrial-control machine does its utmost to repress the ecological potentials of both life and theatre.[1] Taken together, their singular contributions provide a new lexicon for Theatre and Performance Studies, a set of heterogenous co-ordinates for charting the ecology immanent to the theatrical medium.

Barad: diffractive ecologies

Karen Barad is one of the foremost posthumanist thinkers working today, someone whose ideas on quantum performativity have found a receptive audience amongst Theatre and Performance scholars.[2] Her work is particularly germane to theatre ecology, since like Simondon before her, she liberates agency from its jealously guarded humanist estate. Instead, she aligns it with an anonymous force that inheres in the very stuff of matter, the molecular relations that drive the virtual 'acts' of a quantum universe, which is always performing itself:

> Agency is not aligned with human intentionality or subjectivity. Nor does it merely entail resignification, or other specific kinds of moves within a social geometry of antihumanism. Agency is a matter of intra-acting; it is an enactment, not something that someone or something has [...]. Agency is not an attribute whatsoever – it is 'doing/being' in its intra-activity. (Barad, 2003: 826–7)

What Barad terms 'agential realism' is predicated on intra-activity.[3] As its prefix implies, intra-activity designates a diffractive relationality whereby transformation happens *within* things rather than simply *between* them: in quanta as well as relata. Applying the diffractive experiments of both physics and quantum physics to philosophy, Barad is able to state that reality is neither bounded nor stable. In the same way that a single wave of light diffracts into a pattern of multiple waves when it is passed through a pinhole slit in a sheet of paper or card, so the actual state of the world is never one or whole. It is always haunted by a number of virtual possibilities that are invisible to the naked eye but nonetheless there. Beneath the seeming stability of the world, the virtual is already at play, wagering on the multiple, diffracting the one:[4]

> Diffraction is not a set pattern, but rather an iterative (re)configuring of patterns of differentiating-entangling [...]. Matter is itself diffracted, dispersed, threaded through with materializing and sedimented effects of iterative reconfigurings of spacetimemattering, traces of what might yet (have) happen(ed). Sedimenting does not entail closure. (Mountain ranges in their liveness attest to this fact). (Barad, 2014: 168)

A key element in Barad's quantum notion of diffraction stems from her interest in the double split experiment in modern physics. In that experiment, light is seen to be both a particle and a wave at the same time. Matter, then, is fissured from the beginning, haunted by itself. Even when it is one, it is other.[5] A decidedly queer state of affairs:

> Electrons are queer particles, '*mita y mita*'. They are particles. They are waves. Neither one nor the other. A strange doubling. A queer experimental finding. A theoretical impossibility (at least from the point of view of classical Newtonian physics). Unable to account for its inappropriate behaviour, physicists label it 'wave-particle duality', a disturbing paradox. (Barad, 2014: 173)

In Barad's diffractive model of physics, based on the quantum thinking of Niels Bohr, a new ethics is required; one that is no longer solely concerned with confining the actual, pinning it down and extracting from it, as capitalist economics always wants to do. Rather, to be ethical is to be open to the diffractive excess – the 'paroxysm' (Barad, 2010: 245) – which eludes capture by the intentionality of the humanist subject.[6] Ethics is a matter of potential, being open to as many futures as possible:

> A delicate issue of ethicality runs through the marrow of being. There is no getting away from ethics – mattering is an integral part of the ontology of the world in its dynamic presencing. (Barad, 2007: 396)

In the language of this book, earthly life is ethical for Barad because it is *theatrical*, a splitting or diffractive relation in which objects and bodies diverge from themselves and elude synthesis.[7] Barad's usage of the word 'presencing' is resonant in this context. Like Alan Read's notion of 'arriving' (see pp. 47–8), presencing assumes that actors never coincide with their role. Presencing is theatrical because, in it, endings are provisional and perpetually deferred – the actor gets up after feigning their death; the show starts again the next night; a *Hamlet* performed in 1605 is very different from one performed in 2023. In the theatre, reality is heterogeneous, constantly diffracting from a 'source text' that nevertheless haunts it, provoking what we could call, after Barad 'spooky entanglements' – a throng of ghosts that collapse distinctions between past, present and future, and which gesture towards the excess that every theatre production possesses.[8] No actual staging will ever exhaust the virtual possibility of a play. The last word can never be said. There is always something to come, a new context that changes the original, an indeterminacy that is foundational.[9]

In the respect to which Barad's 'spookiness' is located at the very 'heart of matter itself' (Barad, 2010: 251, 249), her ideas affect a veritable revolution in *both* ecological ethics *and* theatrical praxis.[10] Instead of just dealing with actuality, Barad asks us to commit to the virtual force in the theatrical event that no concrete image or signifier can hope to figure or contain:

> In fact, this indeterminacy is responsible not only for the void not being nothing (while not being something), but it may in fact be the source of all that is – a womb that births existence [...]. *The vacuum is far from empty; rather it is flush with yearning, with innumerable possibilities/imaginings of what was, could be, might yet have been, all co-existing.* (Barad, 2018: 232; original italics)

Against all committed dramatic discourses, such as say, Bertolt Brecht's, that would argue for visibility, the showing of appearance, in the name of a 'clear social function' to come (Brecht, 1964: 128), Barad's quantum notion of diffraction allows one to see that theatre's ecological power resides in its ontological capacity to de-actualise, de-realise and un-do: to shower the world with the ungraspable power of the virtual. Differently from the Brechtian performer whose role in the celebrated 'street scene' of epic theatre is merely analogous to reality, Barad's performer has already brought new, possible worlds into being by the simple fact that they have stepped on to the stage and committed to the realm of the theatrical.[11] As a 'single event that is not one' (Barad, 2010: 244), theatre makes diffraction palpable; it explodes the very idea that a single reality or identity exists:

> Diffraction is not a singular event that happens in space and time; rather it is a dynamism that is integral to spacetimemattering. Diffractions are untimely. Time is out of joint; it is diffracted, broken apart in different directions, non-contemporaneous with itself. Each moment is an infinite multiplicity. (Barad, 2014: 169)

Barad's contribution to theatre ecology is radical and profound. By insisting on diffraction, a 'fracturing' within matter that resists repair, she interrogates the efficacy of commonsensical models of ecodramaturgy that would supposedly impact on some 'real' world existing beyond the walls of the playhouse. For if the real is not a phenomenal thing that stands apart from its representations in a discrete pocket of existence, but a set of relations that are always under construction, then it makes no sense to want to reflect the real. Such an approach only petrifies the real, keepings things as they are, shutting down the possibility of life to be other than what it is. Rather, the imperative is to bring actuality to crisis, getting to the point where the virtual is released like a kind of 'plague' or virus, to pre-empt the terms of Artaud. On the ecological stage, then, what are needed are not explicit representations of the environment but diffractive performances that unfix naturalised images and identities, that theatricalise the spacetimematter of the real. By doing so, the possibility of a new world is made palpable in the here and now. That world does not need to be transferred from the stage to another place. There are no insides and outsides. Only one reality, in which everything impacts on everything else. All of this, of course, is already inherent to theatre, but what Barad does is to provide the theatre ecologist with a lexicon of new terms – 'diffraction', 'virtuality', 'intra-activity', 'spooky entanglements' – for describing the ecological effects that theatre can have when it insists, as much as it can, on being itself.

In the wake of Barad's thought, theatre becomes an intra-active event that troubles the real, producing ecological possibilities in productions that, on the surface, might appear to have only a tenuous relationship with environmentalism. One thinks, for instance, of the haunting absence of 'Nature' in Beckett's work, or the strange, sci-fi plays of Alistair MacDowell or the uncanny cyborg worlds created by Tadeusz Kantor. For these artists, theatre is a space that de-creates the real, offering virtual alternatives to the dystopian, artificial landscapes that their work represents in terms of content and setting – the plane of the actual. Regardless, however, of her extensive predilection for performative metaphors, Barad makes no attempt to make an aesthetic leap into the messy and fleshy business of theatre practice *per se*. Like so many posthumanist ecologists (see pp. 91–5 in Chapter 2), her default position is always to read performance, as if intra-activity was something that, despite its materialist ontology, could only ever be played out on a page, in a forest of signs. On those rare occasions when Barad does talk of theatre – I am thinking of her multiple references to Michael Frayn's 1998 play *Copenhagen* in her book *Meeting the Universe Halfway: Quantum Physics and the Entanglement of Matter and Meaning* (Barad, 2007) and the essay 'Quantum Entanglements' (Barad, 2010) – there is no study of any actual mise-en-scène. Rather, she is content to interpret performance as a textual artefact, an object that serves an already predetermined project. While certainly pertinent theoretically, Barad's ideas on virtuality only get theatre ecology so far. To widen the ecotheatrical potential of the virtual, to make it *really* matter, greater attention has to be paid to theatre's affective power. This is all the more pertinent since the affects that certain theatres generate are often more existentially unsettling and uncomfortable than the largely optimistic and jubilatory ones that Barad tends to associate with agential realism. As a way of reminding oneself of the basal, sensate dimension of the theatrical encounter, it is useful to turn to the eco-aesthetic thinking of Jean-François Lyotard, someone who approaches western theatre as a sensorium, a 'more-than-verbal mechanism' for intervening into the world.[12]

Lyotard: *oikos*

In his essay '*Oikos*', written in 1988 on the cusp of the 1992 Rio Earth Summit, the first collective response to climate change made by world governments, Lyotard argues that conventional, scientific understandings of ecology fail to address the 'situation of distress, of suffering' that human subjects feel in the fact of living (Lyotard, 1993a: 106). For Lyotard, humans are anguished beings, weak and belated creatures whose 'psychic apparatus', forged in a 'phase that precedes language', is constantly under threat from the presence of a disruptive, non-human 'guest' – an *oikos* – that language can 'touch' but never capture. As Lyotard has it, there is something about this guest, this *oikos*, that posits ecology as an exercise in existential suffering, a coming to terms with an impossible, ineradicable doubleness, an 'apartness':[13] 'I mean simply, that for me, "ecology" is the discourse of the secluded, of the thing that has not become public, that has not become communicational, that has not become systemic, and that can never become any of these things' (Lyotard, 1993a: 105).

If ecology (*oikeion*), Lyotard opines, is to have any meaning at all, if humans are to flourish on an agentic earth, the dominant environmental lexicon (sustainability, conservation, resilience, stewardship) will need to be rethought, perhaps even abandoned. Against all extant environmental normativity, the better course of action, Lyotard suggests, provocatively, is to unwork one's identity. Lyotard urges us to leave the non-human guest (the *oikos*) in its shadowland, not to shine light on it or to make it productive. The gnawing presence of this terrestrial thing, this stranger at the heart of the human, 'presupposes', Lyotard says, 'that there is a relation of language with the logos, which is not centered on optimal performance and which is not obsessed by it, but which listens to and seeks for what is secluded, *oikeion*' (Lyotard, 1993a: 105).

Given the 'billion black anthropocenes' that so many of the poor, disenfranchised and Indigenous multitudes are already living through today, the ecological provocation suggested by Lyotard could appear disingenuous, offensive, even, the pure thought of a white soul. But such a critique needs to be resisted. For to mobilise the *oikos*, to put it to use, is to repeat the logic of capitalist imperialism that has lost all faith in the earth and the peoples who inhabit it. As Glissant has it, the capitalist coloniser looks to transform 'opacity' into 'transparency' (Glissant, 1997: 62). The *oikos*, for Lyotard, is opposed to that transparency. It demands a different economics, a new ratio, an alternative mode of being. Never can the *oikos* be equated with any kind of measure, something owned or made readily available:

> When *oikos* gives way to *oikonomikos* or *oikonomikon*, a complex transformation of the word *oikos* itself occurs. If 'economic' means *öffentlich*, it implies that the *oikos* itself has slipped away elsewhere. (Lyotard, 1993a: 105)

Ecology, for Lyotard, is sensate and terrestrial, a somatic existentialism haunted by 'the question of birth', the fact that we are material creatures, destined to die (Lyotard, 1993a: 107).[14] As Lyotard has it, there is no redemption in matter, no messianic transcendence that would exist beyond the world, as Barad's references to ghosts and spectres sometimes seem to imply.[15] By contrast, in the face of radical immanence, there is a requirement to admit anthropocentric defeat, to give into loss, to embrace suffering. For without an acceptance of discomfort, there is no possibility of believing in the earth as earth. 'The body and the mind', Lyotard insists, 'have to be free of burdens' for ecology to exist. 'That doesn't happen without suffering. An enjoyment of what we possessed is now lost' (Lyotard, 1999: 19).

One of the ways that this eco-existentialist suffering can be mitigated, if not ever overcome, Lyotard suggests, is through the power of aesthetics, the capacity of the artwork to stimulate corporeal feelings and lively emotions. Hence Lyotard's love of theatre, the medium he lauds as a 'somotography', a staging of and writing with bodies. Crucially, theatrical bodies are not required to make semantic sense, for Lyotard; on the contrary, their purpose is to disfigure meaning, producing in the process 'a general dissemioticization', energy for liberation (Lyotard, 1997a: 285):

> The business of an energetic theatre is not to make allusion to an aching tooth when a clenched fist is the point, nor the reverse. Its business is neither to suggest that such and such means such and such, nor to say it, as Brecht wanted. Its business is to produce the highest intensity (by excess or by lack of energy) of what there is, without intention. (Lyotard, 1997a: 288)

Conventional binary distinctions between theatres of hope and despair lose their relevance when it is a matter of staging the *oikos*. Instead, what is required of spectators is *passibility*, a willingness to undergo 'the miraculous but precarious condition' of the performance, to open oneself to an experience that undoes the habitual gap between passivity and activity (Lyotard, 1997b: 245). For Lyotard, to be *passible* is to be made receptive, a spectator capable of affirming 'a form of life that was spiritual because human, human because earthly – coming from the earth of the most living of living things' (Lyotard, 1991: 9). Through the exclusions and binaries of capitalist modernity, modern human beings have been historically conditioned to repress all knowledge of this *oikological* life. Disastrously, the emphasis has been put on volition and autonomous action, in making the world available, calculable, ripe for exchange. However, as we know from recent psychoanalytical work on climate trauma, the more the autonomy of the earth is denied, the greater the violence inflicted on bodies and minds, the more intense the derangement.[16] By allowing the *oikos* a space to express (if never to show) itself, Lyotard's theory of theatre wagers on the 'lesser violence', an affective intensification or 'cruelty' that decentres the human. In this corporeal *oikology*, the earth does not exist uniquely for humans to use and neither are we called upon to save it. Rather it deserves to be affirmed, even when it cares little for us. For, as Lyotard has it, terrestrial existence is a continual awakening, a fragile, vulnerable becoming that is open to the violence of matter, a force that inflicts wounds on bounded identities:

> Existing is to be awoken from the nothingness of disaffection by something sensible over there. An affective cloud lifts at that moment and deploys its nuance for a moment [...]. What we call life proceeds from a violence exerted from the outside on a lethargy. (Lyotard, 1997b: 243; original italics)

By introducing concepts such as '*oikos*', '*passibility*', and 'energetics' into Theatre and Performance Studies, Lyotard adds to the lexicon that Barad provides for grasping the ecological doubleness of theatre, the way in which the non-human is immanent to the medium as a force that provokes becomings. But more productively than Barad, he is also sensitive to the corporeal aesthetics of the performance event, a point that I will unpack in greater detail in Chapter 6 when I discuss his concept of the *gestus*. Yet despite these generative additions, as with Barad, there are limits to Lyotard's insights. Something else is required if theatre ecology is to fulfil its potential: a conceptual bridge that would explain, in concrete terms, how eco-aesthetic experience is translated into collective experience, thus producing a political becoming. To construct that link, I turn now to the transversal thinking of the philosopher and clinician Félix Guattari, someone for whom a progressive ecology is found not in saving 'Nature' but in demanding a fundamental overhaul in the very notion of human subjectivity itself.

Guattari: transversal subjectivity

More explicitly than either Barad or Lyotard, Guattari is committed to creating new forms of ecological being. The goal in his clinic is to opt for life rather than death, to cultivate a 'sense of responsibility for the future of all life on the planet, for animal and vegetable species, likewise for incorporeal species such as music, the arts, cinema' (Guattari, 1995: 120). Appropriately for an ecological thinker, Guattari's environmentalism is not content to argue for any kind of discrete action. In a world dominated by the hegemony of international world capitalism's 'sad passions', ecological transformation presupposes nothing less than a new milieu for living, a total recalibration of what it means to think and feel as a human subject (Guattari, 2000: 32–4):[17]

> The ecological crisis can be traced to a more general crisis of the social, political and existential. The problem involves a type of revolution of mentalities whereby they cease investing in a certain kind of development, based on a productivism that has lost all human finality. (Guattari, 1995: 119)

To bring about such a profound revolution in thought, Guattari looks to rethink subjectivity transversally, as an interface or platform in which a number of competing discourses, forces and values momentarily align and find a fragile consistency.[18]

> The subject is not a straightforward matter; it is not sufficient to think in order to be, as Descartes declares, since all sorts of other ways of existing have already established themselves outside consciousness. [...] Rather than speak of the 'subject', we should perhaps speak of *components of subjectification*, each working more or less on its own. (Guattari, 2000: 24–5; original italics)

Unlike capitalist subjectivity, transversal subjectivity is neither fixed nor identified with an individual self. It is more accurately approached in non-human terms as territory that is being constantly modified by the contexts in which it finds itself as well as the information, social codes and flows of desire that cut across and through it. And it is precisely because subjectivity is constituted through the transversal interplay of multiple components that Guattari attaches such hope to it. For when one of these components starts to break down or deterritorialise itself, the 'fall out' impacts on everything else, allowing for new 'mutant' universes of value to come into being in places well beyond the initial point of contact (Guattari, 1995: 120).

Some indication of how this 'mutant creativity' might be brought about practically is provided by looking at how Guattari organised his own therapeutic sessions. At La Borde, a clinic he ran with Jean Oury in the Loire valley, special attention was accorded to the psychoanalytical idea of transference, the process whereby the analysand identifies with something enigmatic in the analyst – Jacques Lacan calls this the '*petit objet a*' – that allows the patient to extricate themselves from the cycle of despair and depression in which they are caught.[19] Typically, however, Guattari is not satisfied with orthodox practices of psychiatry. Rather, influenced by the ideas of the literary critic Mikhail Bahktin, he moves transference beyond the privatised space of the psychoanalytical clinic and situates it, instead, within the 'contact zone' of aesthetic experience.[20]

Following Bakhtin, Guattari insists that eco-aesthetic transference unfolds in a triple movement. Initially, it starts with artists capturing materials from everyday life and infusing their content with expressive energy; in the second stage, artists translate that expressiveness into material forms that stand on their own; and, in the third, spectators engage with the work and find themselves mixed up in and changed by it. By 'agglomerating' to artworks, 'something', Guattari says, 'is detached and starts to work for itself and for you' (Guattari, 1995: 132–3):

> The assemblages of aesthetic desire and the operators of virtual ecology are not entities that can be easily circumscribed within the logic of discursive sets. [...] They are becomings – understood as nuclei of differentiation – anchored at the heart of each domain, but also between the different domains in order to accentuate their heterogeneity. A becoming child (for example in the music of Schumann) extracts childhood memories so as to embody a perpetual present which installs itself like a branching, a play of bifurcations between becoming women, becoming plant, becoming cosmos, becoming melodic. (Guattari, 1995: 92)

In Guattari's theory of aesthetico-existential production, the virtual ecology of the work is not represented as an idea or theme to think through hermeneutically. Aesthetic transference, as Guattari has it, is sensate and corporeal, an affective bifurcation that severs and sutures. It spills over and recreates the subject, affording new worlds, suffusing the world with a kind of magic. Transference is an animating force, provoking the de-phasings necessary for life to evolve:

> Art confers a function of sense and alterity to a subset of the perceived world. The consequence of this quasi-animistic speech effect of a work of art is that the subjectivity of the artist and the 'consumer' is reshaped. [...] The work of art for those who use it, is an activity of unframing, of rupturing sense, of baroque proliferation, which leads to recreation and reinvention of the subject itself [...] The event of its encounter can irreversibly date the course of an existence and generate fields of the possible 'far from the equilibria' of everyday life. (Guattari, 1995: 131)

Although Guattari privileges the generative power of the artwork, it is important to note that creativity is not limited to human agency. Rather, as I touched on in my explanation of Deleuze's and Guattari's earth in the Introduction (see pp. 13–17), artistic becoming is an ontological given, a property of the planet itself, a drive inherent to all of life, be that organic or inorganic, to express itself, to make connections and leaps. In proximity to such an expansive aesthetics of life, there is no need to depend on environmental politics, governmental policy or activist engagement to catalyse ecological change. Instead, one can start anywhere, at any time, in all those places where experimentation is in operation:

> What is important to know is if a work leads effectively to a mutant production of enunciation. The focus of artistic activity always remains to a surplus-value of subjectivity or, in other terms, the bringing to light of a negentropy at the heart of the banality of the environment – the consistency of subjectivity only being maintained by self-renewal through a minimal, individual, or collective resingularisation. (Guattari, 1995: 131)

By refusing to separate aesthetic experience from either environmental or political thinking, Guattari discloses where the alternative activism of theatre resides. This is found not in representing what spectators already know, but in encouraging them to become other than what they are, affecting them at the very heart of their subjectivity, the experiential interface that brings differences into contact. Two citations highlight how and where Guattari's ideas function within the crucible of the theatre event itself. In the first, he describes performance art as a dizzying event that defamiliarises the present though the production of 'more-than-linguistic' intensity:

> Performance art delivers the instant to the vertigo of the emergence of Universes that are simultaneously strange and familiar. It has the advantage of drawing out the full implications of this extraction of intensive, a-temporal,

> a-spatial, a-signifying dimensions from the semiotic net of quotianity. It shoves our noses up against the genesis of being and forms before they get a foothold in dominant redundancies [...]. (Guattari, 1995: 90)

And, in the second, he contends that the proto-Absurdist theatre of Stanislaw Witkiewicz produces 'new universes of subjectivation' through a deliberate cultivation of strangeness:

> One need only evoke the desperate quest of Witkiewicz to grasp an ultimate strangeness of being, which literally appeared to slip between his fingers. In these conditions, the task of the poetic function, in an enlarged sense, is to recompose artificially rareified, resingularised universes of subjectivation. For them, it's not a matter of transmitting messages, investing images as aids to identification, patterns of behaviours as props for modelisation procedures, but of catalysing existential operators capable of acquiring consistence and persistence. (Guattari, 1995: 19)

In these citations, Guattari specifies where theatre's concrete contribution to ecology resides. In both instances, it is found in the capacity that living bodies have to tap into and express the very 'genesis' of a world before it has been disciplined and mastered by dominant discourses, be they ethical, political or narrowly environmental. By choreographing human and non-human bodies that vibrate with wordless energies, theatre and performance are able to provoke vertiginous ecologies. Where Barad and Lyotard remain somewhat vague about how the political implications of their eco-aesthetic thinking play out and to whom they are addressed, Guattari's is unambiguously militant. The goal of performance, he exhorts, is to reforge world and earth by activating the transversal line that would bring subjectivity, the social and the environmental into temporary alignment – the production of affinities between what he calls the 'three ecologies' of mind, politics and nature. These transversal affinities are the types of experience theatre ecology should be looking to produce:

> The refoundation of politics will have to pass through the aesthetic and analytical dimensions implied in the three ecologies: the environment, the *socius* and the psyche. We cannot conceive of solutions to the poisoning of the atmosphere and to global warming due to the greenhouse effect, or to the problem of population control, without a mutation of mentality, without promoting a new art of living in society. We cannot conceive of international discipline in this domain without solving the problem of hunger and hyperinflation in the Third World. We cannot conceive of a collective recomposition of the *socius*, correlative to a resingularisation of subjectivity, without a new way of conceiving political and economic democracies that respect cultural difference – without multiple molecular revolution. [...] The only acceptable finality of human activity is the production of a subjectivity that is auto-enriching in its relation to the world in a continuous fashion. (Guattari, 1995: 20–1)

Conclusion

In this chapter, I have assembled a heterogeneous model for explicating the immanent ecology that theatre already possesses. As I demonstrated, Barad highlights theatre's capacity to produce virtual diffractions; Lyotard shows how its corporeal exchanges bring the distress of the *oikos* into play; and Guattari explains how its affective power can set in motion a transversal pragmatics that cuts across the planes of politics, ethics and environment in order to make a series of unpredictable interventions. But while the concepts of Barad, Lyotard and Guattari provide all of us – critics, spectators, students and practitioners alike – with a lexicon to reconsider our inherited ways of operating ecologically in the theatre, the point is not to see theatre ecology as Baradian, Lyotardian or Guattarian. Rather, in all instances, the aim is to allow access to theatre's own ecological modes of operating. Theatre's embodied lines of flight and sensate geometries cannot be predicated or canalised in advance by an overarching theory, unitary concept or single discourse. Instead, theatre ecology is more productively approached as a constellation, a protean and machinic medium for the creation of new planetary values, sensations and subjectivities. It demands performances that are virtual and transversal, corporeal spacetimes where the *oikos* is touched and brought into play.

Notes

1 Although there is no purposeful attempt to break with the capitalist status quo in his writing, it would nevertheless be a mistake to think that there is a lack of politics to Simondon's thinking. Simondon's ideas on politics pertain to his collectivist-inspired notion of transindividuated being and are grounded in openness, the capacity of individuals and groups to re-create themselves beyond technocratic capture. As Andrew Lapworth contends, Simondon's project was to propose a third way between science and culture in a manner different from C. P. Snow in the UK: 'What defines the novelty of Simondon's response to this conflict is the claim that achieving a potential rebalancing of human and machine is not just a *political economic* matter of transforming production (à la Marx) or reforming our educational institutions (à la Snow), but is instead, first and foremost, a philosophical problem: it concerns our very capacity *to think technology in the reality of its inventive becoming*' (Lapworth, 2020: 109; original italics).
2 See, for instance, Barad's central position in Silvia Battista's *Posthuman Spiritualities in Contemporary Performance: Politics, Ecologies and Perceptions* (Battista, 2018: 16–18, 121–31, 173–80) as well as in the edition of *Performance Research* 'On Diffraction' (2020) dedicated to her work, edited by Annouchka Bayley.

3 Note that it is agential realism, not speculative realism. Matter performs and implicates in the here and now; it produces affects and effects in actual sociopolitical worlds, even if those effects cannot be seen with the naked eye. But, as with all realism, the fact that we are unable to see the agency of the universe does not mean that it does not exist. See Rebekah Sheldon (2015: 193, 214).
4 Barad explains diffraction by providing an account of the experiments of the seventeenth-century scientist Francesco Grimaldi in Bologna. Not only did Grimaldi discover that light diffracts when it is passed through a pinhole on paper, but he discovered that there was both light in darkness and darkness in light: 'Bands of light appear inside the shadow region – the region of would-be total darkness; and bands of darkness appear outside the shadow region. There is no sharp boundary separating the light from the darkness: light appears within the darkness within the light within. [...] Grimaldi is clear that the explanation for these remarkable findings could not lie with the corpuscular theory of light. Imagining light to behave as a fluid which upon encountering an obstacle breaks up and moves outwards in different directions, Grimaldi dubbed this phenomenon diffraction, citing the Latin verb *diffringere* – *dis* (apart) and *frangere* (break)' (Barad, 2014: 170–1).
5 Barad describes the huge ontological and ethical shift that quantum physics entails in the following way: 'According to classical Newtonian physics, the two-slit diffraction apparatus is the ultimate ontological sorting machine – it unambiguously differentiates particles from waves: waves make diffraction patterns because they can go through both slits at once, particles don't. But in the early twentieth century electrons passing through a diffraction apparatus fail to behave like proper particles. Rather they behave like waves. Indeed, it seems that each individual electron is somehow going through both slits at once. (Talk about inappropriate!) To make matters worse, each individual electron arrives at one point on the screen just like a proper particle. Now add a which-slit detector to the apparatus (to watch an electron going through the slits) and the electrons behave like particles. Impossible they say, but this is the electron's lived experience' (Barad, 2014: 173).
6 One finds the word 'paroxysm' in the theatrical lexicons of Jean Racine and Artaud. Interestingly, Barad uses it in an essay on *Hamlet*.
7 I use theatrical rather than performative, in this context, in order to stress the excessive surplus of matter, the sense in which, in any theatrical production, there is always something that escapes the actual, haunting reality with a thousand possibilities that may or may not come into being. The performative tends to be more delimited in its creation of excess, more concerned with being productive, with trying to replicate itself, efficiently. Of course, the performative too, like any autopoetic act, is unable to coincide with its original intention, but its 'noise' (difference) tends to be controlled, re-invested, less wild in the scale and scope of its becomings. Another way of saying this, and this would relate to Jon McKenzie's ambivalence towards the word in *Perform or Else: From Discipline to Performance*, is that the performative has a greater investment in doing rather than undoing, in being utile (McKenzie, 2001: 55, 94).

8 Barad uses the phrase 'quantum entanglements' (Barad, 2010: 241), but she could easily have said 'spooky' entanglements. 'Spooky' is one of her favourite words.
9 No one could ever have thought that the hurricane – Katrina – that hit New Orleans in 2005 would radically change the meaning of Beckett's figure of 'Godot', making it ecopolitical, sedimenting its meaning, making it matter differently, as Paul Chan demonstrated in his site-based performance of the play in 2007.
10 For instance, in the essays 'No Small Matter: Mushroom Clouds, Ecologies of Nothingness, and Strange Topologies of Spacetimemattering' (Barad, 2017) and 'Troubling Time/s and Ecologies of Nothingness: Re-turning, Re-membering, and Facing the Incalculable' (Barad, 2018), Barad is dedicated to teasing out the environmental ethics of agential realism. She does so by concentrating on the ghosts of Hiroshima and Nagasaki.
11 Brechtian actors show reality not as it is, but how it could be interpreted. They do so by acting as if their gestures and words were citational, an overt demonstration of what happened as opposed to some supposedly 'real' imitation of the original event. 'In short', Brecht underlines, 'the actor must remain a demonstrator; he must present the person demonstrated as a stranger' (Brecht, 1964: 125). Unlike Barad, however, Brecht does not question the solidity or unity of the reality the actor cites; rather, he interrogates the veracity of its representation.
12 Speaking of a formative encounter with the New York theatre-maker Richard Foreman in the 1970s, Lyotard says: 'Foreman's advantage over me was that he was convinced of how necessary it is to put intensities on stage and how very little relevant it is to perform their presentation in flesh and blood in the text. There was something outrageous with respect to Being in my attempt to equate it with text. The scandal lay in the claim that writing holds the position, no matter how uncertain with the law. Nothing other than parody could follow from such a claim' (Lyotard, 1988: 14).
13 There are obvious resonances here with Barad's idea of 'spooky entanglements' but the major difference resides in the fact that Lyotard is anguished by an internal ghost, concerned that identity will be overwhelmed by uncanniness, a strangeness from within the psyche.
14 A recurrent trope in Lyotard's ecological thought at the end of the 1980s is the image of an exploded, vaporised earth, the planet as victim of the heat-death of the sun. As part of his 'postmodern fable', Lyotard uses the image to reflect on how, in advanced capitalist societies, a privileged technocratic class is gearing up to leave the earth behind. For the sake of survival, this class are willing to become inhuman(e), data machines whose information can be downloaded to some supercomputer, and whose bodies are forgotten and denied. The tragedy, here, is that bodies, for Lyotard, provide the flesh that links humans to terrestrial existence, the matter which gives life its quality and passion. In the absence of flesh, art, childhood, ethics itself are eradicated: 'No need for writing, childhood, pain. To think consists in contributing to the amelioration of the big monad. It is that which is obsessively demanded of us. You must think in a communicable

way. Make culture. Not think according to the welcome of what comes about, singularly. To pre-vent it, rather. To success is to process [*sic*]. Improve performances. It is a domestication, if you will, but with no *domus*. A physics with no god-nature. An economy in which everything is taken, nothing received. And so necessarily, an illiteracy' (Lyotard, 1991: 199). For an excellent reading of this postmodern fable, see Chapter 1 of Ashley Woodward's *Lyotard and the Inhuman Condition: Reflections on Nihilism, Information and Art* (Woodward, 2016: 11–40).

15 Barad's hauntology is drawn from the thinking of Jacques Derrida and Emmanuel Levinas, both of whom see the subject as being ethically bound to ghosts and specters, the call of the Other who is radically absent. This onus on transcendence explains why Silvia Battista is able to make use of Barad's ideas for new forms of ecological spirituality (Battista, 2018: 4).

16 See work by Joseph Dodds (2011), Simon Estok (2018), Lee Zimmerman (2020). For these thinkers, climate change is existentially traumatic. It creates anxiety, depression and panic and impacts on the unconscious life of subjects, producing strange and sometimes violent behaviours. Consider this from Zimmerman: '[D]ominant discourses of global warming contribute to a form of denial, beyond the overt form associated with the officially designated 'denialists' or 'skeptics', and that we might understand that more fundamental, pervasive and largely unrecognized species of denial by thinking of global warming as a type of trauma' (Zimmermann, 2020: 1).

17 This long citation explains the importance Guattari attaches to a transversal revolution: '[I]t is no longer possible to claim to be opposed to capitalism's power only from the outside, through trade unions and traditional politics. It is equally imperative to confront capitalism's effects in the domain of mental ecology in everyday life: individual, domestic, material, neighbourly, creative or one's personal ethics. Rather than looking for a stupefying and infantalising consensus, it will be a question in the future of cultivating a dissensus and the singular production of existence. A capitalist subjectivity is engendered through operators of all types and sizes, and is manufactured to protect existence from any intrusion of events that might disturb or disrupt public opinion. It demands that all singularity must be either evaded or crushed in specialist apparatuses and frames of reference. Therefore it endeavours to manage the worlds of childhood, love, art, as well as everything associated with anxiety, madness, pain, death, of a feeling of being lost in the Cosmos […]' (Guattari, 2000: 33).

18 Initially, Guattari talked of the transversal as a clinical term that would replace the emphasis that Freudian and Lacanian models of psychoanalysis placed on the transference. Where the classic transference takes place between analyst and analysand on a bourgeois couch, the transversal was intended to take into account collective modes of therapy, institutional contexts and alternative ways of organising a patient's recovery. The point, as always in Guattari's thought, was to find moments of ecological connection, conjugation and transformation between differences without creating synthesis or unity. For more on the evolution of this concept in Guattari's work, see his *Psychoanalysis and*

Transversality: Texts and Interviews 1955–1971 (Guattari, 2015) and Gary Genosko's essay 'The Life and Work of Félix Guattari: From Transversality to Ecosophy' (Genosko, 2000: 46–78).

19 In Lacanian therapy, the *'petit objet a'* is not an object at all. It is an enigmatic something, a relation, a force, a voice, a shimmer that sets desire in motion.

20 In Lacan's thinking, the *'petit objet a'* is linked to interstitial parts of the body (mouth, anus, ears etc.). However, anything can be a *'petit objet a'*, for Guattari – a building, music, a gesture, a word, a rhythm, a performance.

4

Concept (ecologising theatre)

Introduction

In this chapter, I look to unfold in greater detail what has been merely assumed in Chapters 2 and 3: namely, that theatre's ecology is immanent to its specifically theatrical way of representing the world. In doing so, I reverse the trajectory taken in Chapter 2 by *ecologising* theatre as opposed to *theatricalising* ecology. My attempt to ecologise theatre in this chapter is composed of two parts. In the first, I return to the Ur-figure of western theatre theory – Plato's cave – in order to acknowledge the ecological potential that theatre has harboured from the very beginning of its theorisation in the West; in the second, I concretise the ideas of Barad, Lyotard and Guattari by tracing the specific ways in which the 'theatricality' of the theatre event provokes an affective, non-human ecology by allowing bodies to reverberate with pre-individual forces and virtual possibilities. Echoing what I argued for at the end of Chapter 1, to think the ecology of theatre is not to propose a genre of environmental performance, it is to interrogate the historical ground and practice of western theatre itself.

Theatre's lithic ecology: re-turning to the cave

As almost every encyclopaedic entry or history of the medium proclaims, theatre is the anthropocentric art form par excellence: the 'machine' whose longstanding commitment has been to showing the actions of human agents, even when their worlds are ruled by gods.[1] The words of the Chorus in *Antigone* are exemplary, in this regard:

> CHORUS: Is there anything more wonderful on earth,
> Our marvellous planet,
> Than the miracle of man!
> With what arrogant ease

> He rides the dangerous seas,
> From the waves towering summit
> To the yawning trough beneath.
> The earth mother herself, before time began,
> The oldest of the ageless gods,
> Learned to endure his driving plough,
> Turning the earth and breaking the clods
> Till by the sweat of his brow
> She yielded up her fruitfulness.
>
> (Sophocles, 1986: 147)

Irrespective of Martin Heidegger's ironic reading of this passage, the cultural logic of modern western theatre has largely remained committed to staging this Sophoclean 'miracle'.[2] From Gotthold Lessing's notion of bourgeois dramaturgy in the mid-eighteenth century through to contemporary experiments with performance lectures, theatre has been habitually portrayed as a night school in humanism, a mimetic medium that teaches spectators what 'human nature' is and to whom its privileges should be granted. As Hana Worthen puts it:

> [W]hen appropriately disciplined to the evocation of individual freedom of expression, theatre is an extension of human nature through which we continually regulate that nature, through the simultaneously curative and constrictive power of mimesis. Theatre, in this sense, provides the necessary means for the realization of human nature, the means to make visible and cognizable what would otherwise remain invisible and unseen to *us*; it provides the necessary self-regulating mimetic instrument, the liberating mirror in which *we-I* = *we-see* what otherwise would remain *in us* and *for us* unseen. (Worthen, 2020: 31; original italics)

While theatre's attempt to make visible a universalist notion of 'human nature' has been rightly critiqued by the performative turn that took place in Theatre and Performance Studies in the 1980s, what has been rarely questioned, at least until recently, is the idea that the stage should be anything other than a crucible for the display and reproduction of human identity alone. This is not to say that dissenting voices have been lacking. It is simply to point out that those critics who did extend their gaze beyond the human – Richard Schechner (1985, 1995), Joseph Roach (1993), Alan Read (1994), Jean-Marie Pradier (2000), David Williams (2000), Jane Goodall (2002) and Nicholas Ridout (2006) – were very much in the minority in Theatre Studies. And even in their writings, the focus has generally been on celebrating theatre's power to include the performances of other life forms and creatures, as opposed to interrogating the humanness of the stage *per se*.[3] The great irony in this long history of forgetting is that a very different, non-human origin for theatre was already in operation from the very beginning of western dramatic theory in Plato's writings on the medium, albeit in obscured fashion.

In his allegory of the cave in Book VII of *The Republic*, Plato describes theatre as prehistoric technology, an assemblage of stone and fire in which spectators watch shadows in an underground den:[4]

> And now, I said, let me show in a figure how far our nature is enlightened or unenlightened: – Behold! human beings living in a underground den, which has a mouth open towards the light and reaching all along the den; here they have been from their childhood, and have their legs and necks chained so that they cannot move, and can only see before them, being prevented by the chains from turning round their heads. Above and behind them a fire is blazing at a distance, and between the fire and the prisoners there is a raised way; and you will see, if you look, a low wall built along the way, like the screen which marionette players have in front of them, over which they show the puppets. (Project Gutenberg.org., 1998)

Plato's elementalist take on theatre inadvertently recalibrates the history, ontology and ecology of the medium. In its essence, Plato shows that theatre is not an affair of anthropocentric signs and affections, as so many western scholars, taking their cue from Aristotle, have professed for so long. Rather, it is a sensorium of 'vibrant matter' (Bennett, 2010) that makes expressive use of cosmic energies and terrestrial substances in order to create scenographic images. Habitually, these 'images' have been understood by scholars of Plato as distorting simulacra, smudged representations that cause spectators to avert their gaze from the clear light – the truth – of the sun lying beyond the dark hollow of the cave. However, they disturb Plato for additional reasons. As well as providing spectators with a faulty epistemology, a wrong way of looking, theatrical images tether them to the earth, interpellating them as creatures of sensation that are bifurcated and changed through their encounters with base matter. In other words, theatricality is mistrusted by Plato not just because it confuses 'substance with shadow', but also because it materialises substance. Because of its interest in variation and transformation, theatre shows the world to be subject to what Simondon would see as the 'de-phasings' and 'individuations' of a planet that is never still, bringing the sky-God eternity of the Platonic idea down to earth.

This alternative way of thinking through theatre's 'geo-scenography' is present again in the anxiety that Plato finds in the affective spell that the actor or imitative poet is able to put on spectators. As Plato has it, the performer spreads 'dis-ease' by using representation to unleash an excess in human and non-human bodies – both a pathos and a virtuality – that spectators are drawn to and seduced by. For Plato, the performer's production of affect is a type of 'evil', an unpredictable, non-human force that queers identity and disrupts all ratio and reason:

> As in a city when the evil are permitted to have authority and the good are put out of the way, so in the soul of man, as we maintain, the imitative poet implants an evil constitution, for he indulges the irrational nature which has no discernment of greater and less, but thinks the same thing at one time great and at another small – he is a manufacturer of images and is very far removed from the truth. (Project Gutenberg. org., 1998)

Undermined by the duplicitous status of the performer, as well as by the geolithic images of the stage (cave), the great chain of being loses its stability. Positions and taxonomies are blurred; and the human subject is decentred and orphaned. In theatre, the ontological scenario that terrifies Plato is embodied. Chaos not only rules; it is the catalyst for a different ontology. One in which the two realms of mind and matter, which Plato is so concerned to separate, now co-exist as singular parts of a common, monadic substance. The horror that mixity provokes in Plato finds its textual correlate – its symptomology – towards the beginning of Book X of *The Republic* when Socrates asks Glaucon to consider artistic or theatrical work as akin to the illusory working of a mirror. With each turn of its reflective surface, the mirror fragments the neat world of completed essences and separate substances, unleashing a non-human plague of images and representations, distancing the human from its proper place in the world:

> For this is he [the artist/mirror man] who is able to make not only vessels of every kind, but plants and animals, himself and all other things – the earth and heaven, and the things which are in heaven or under the earth; he makes the gods also. (Project Gutenberg.org., 1998)

A standard, performative interpretation of this passage, at least one that wants to read Plato 'against' the grain for political purposes, would doubtless concentrate on how the 'mirror' shows the world to be a construct open to change. Yet, that is not the path I want to take. Read ecologically, what is pertinent is not simply that the mirror artificially 'doubles' the world through reflection, but that in the moment of its turning, everything – suns, earths, animals, plants – is spliced together, part of a dedifferentiated sheet of matter, in which multiplicity and sameness co-exist within each other.[5] To turn a 'mirror round and round' is to come into contact with a whorl of stuff, a kind of dynamic formlessness that, as Robert Smithson playfully pointed out in his mock-heroic essay *A Tour of the Monuments of Passaic, New Jersey* (1967), determines being as molecular, prey to the destructive-creative forces of entropy, the irreversibility that brings variation into the world and permits the new to emerge.[6] As this ecomaterialist rereading of Glaucon's mirror reinforces, theatre is to be avoided, for Plato, not simply because of mimetic reasons, but also because of what, in reference to Lyotard, might be termed its 'matter richness', its agentic capacity to shapeshift and take on a different form.[7]

At the level of morphogenesis (the process that creates forms and structures), theatre is dangerous, for Plato, because it is an apparatus that uproots idealist ideas of form and substance and replaces them with dynamic, tempo-terrestrial processes that humans are unable to extricate themselves from or completely harness. That this morphological anxiety in Plato is not a minor issue, something merely epiphenomenal to larger debates about political representation, is underlined, yet again, in his criticism of rhapsodic acting in the dialogue *Ion* – a text in which the elemental qualities that Plato recoiled from in the cave are now transposed to and channelled by the body of the titular performer.[8] As Socrates is careful to point out, Ion's 'gift', like that of Homer's, is not an art or technique that can be learnt and passed on as a discourse; it is an inorganic 'inspiration', an im/material power analogous to the magnetism possessed by the 'stone of Heraclea', the capacity of geological matter to emit a magical charge, to 'touch' without touching, to animate, in other words:

> The gift which you possess of speaking excellently about Homer is not an art, but, as I was just saying, an inspiration; there is a divinity moving you, like that contained in the stone which Euripides calls a magnet, but which is commonly known as the stone of Heraclea. This stone not only attracts iron rings, but also imparts to them a similar power of attracting other rings; and sometimes you may see a number of pieces of iron and rings suspended from one another so as to form quite a long chain; and all of them derive their power of suspension from the original stone. In like manner the Muse first of all inspires men herself; and from these inspired persons a chain of other persons is suspended, who take the inspiration. (Project Gutenberg.org., 1999)

In the same way that the lithic world of the cave hosts, stores and transmits transformative energies, so the 'inspired body' of the actor Ion reverberates with elemental forces. Like the last link in a chain of iron filings or belt of ionised metallic rings, spectators are seduced by the magnetic power of the performer in front of them. At some profound mineral level in their being, and contrary to their will and volition, they tremble in sympathy with the non-human affects channelled through the medium of performing bodies. The 'rhapsody' – the ecstatic music – that Socrates is so suspicious of as he watches the actor, Ion, is not a human song; rather, it emanates from an animated earth, a song that has a more prior claim on the human than knowing and understanding ever could. As ever with Plato, the song is mistrusted because it implicates humans in a world of stones, bones and caves, tying them to a material life that is infused with agency, the metastable performance of a diffractive planet – a pre-individual double – that shatters the fixed essences of solar ideas, placing all bodies on a transversal line, a planetary refrain:

> Do you know that the spectator is the last of the rings which, as I am saying, receive the power of the original magnet from one another? The rhapsode like yourself and the actor are intermediate links, and the poet himself is the first of them. Through all these the God sways the souls of men in any direction which he pleases, and makes one man hang down from another. Thus there is a vast chain of dancers and masters and undermasters of choruses, who are suspended, as if from the stone, at the side of the rings which hang down from the Muse. And every poet has some Muse from whom he is suspended, and by whom he is said to be possessed, which is nearly the same thing; for he is taken hold of. (Project Gutenberg.org., 1999)

To read Plato in spite of himself, to bring him down to earth with the help of the terminology I introduced in Chapters 2 and 3, is to understand that theatre has never been the sole preserve of the human, despite what has been claimed with such persistence for so long. Perversely, theatre is the art form that perhaps best embodies what one could call, after Colebrook, the 'non-humanness of the human', a medium whose mode of signifying troubles, in fundamental fashion, those rigid frontiers separating humans from the earth, matter and spirit, inside and outside, life and death. In the theatre, the human being, like all organic and inorganic matter, is swept up and away by a movement of deterritorialisation. In its very core, at its origin – and this, of course, is what disquiets Plato – the elementalism of the stage forecloses what Colebrook would see as humanism's primary credo: namely, the impossible dream of the human to be substantial to itself, enclosed, uncorrupted, a 'subject of erasure'. What is radical about Plato, the counter-intuitive insight that makes him so much more valuable than Aristotle for ecotheorists and practitioners today, is that he is intensely aware that theatre's power does not derive from the stories it tells. A more primal layer of signification needs to be considered, an excavation that would centre, first, on the stuff it is made from and, second, on its modes of corporeal transmission. In the spacetimematter of the cave, through the 'magnetic pull' of Ion's body, the human is (dis)figured as a terrestrial figure, a biochemical organism that is open to pre-individual transformations and metamorphoses that 'haunt' it with diffractive possibilities. Just like everything else on earth.

Over the past three decades or so, and drawing on important developments in new media, there has been a concerted effort to experiment with the 'more-than-human' technology that Plato apprehended in rudimentary form in the rocks and fire of the cave. In the main, this search has centred on a desire, amongst scenographers and directors, to create a posthumanist, intermedial mode of performance in which animals, puppets, computers, machines and plants have all been enlisted to trouble the boundaries of the human in the very space that is supposed to guarantee its full presence. One thinks of the Wooster Group's screens and monitors; Romeo Castellucci's horses and monkeys; Philippe Quesne's 'moles', DIY gadgets and cars; Kris Verdonck's exploding sculptures and 'inflatables'; Alain Platel's dogs and dodgems; Hirata Oriza's robots; Gisèle Vienne's mannequins and models; and Manuela Infante's 'gardens'. And yet despite these well-known, often-cited examples, there is, as Plato knows, no real need for theatre to experiment, empirically, with non-human things and organisms in its attempt to create what Michel Corvin calls an 'abhumanist stage' in his important but little discussed book *L'Homme en trop: l'abhumanisme dans le théâtre contemporain* (Corvin, 2014) (for more on Corvin, see Chapter 5). All that is required is a recalibration of the gaze. Nothing much, a tilt of the head will often suffice, so that the spectator now pays attention to what was always already so brazenly there within the human performer as a kind of 'purloined letter':[9] the pre-individual or virtual force that shows itself every time an actor leaves the everyday world to appear on the *plateau* (stage),[10] a word whose French etymology, as I have already hinted at (see p. 65), resonates so serendipitously and geolithically with images of tectonic plates and continental drift.

To approach theatre as a terrestrial technology, a cave, is to realise that theatre's ecology is not found just in theme, form and character – the Aristotelian or humanist logic that so many in the Environmental Humanities seem so strangely committed to (see pp. 91–5). Rather, its more powerful manifestation is located immanently, in what one might term the infrastructure of the medium, an expressive, hidden seam in 'theatre's ontology' that troubles the image of the human as a subject rooted in self-coincidence. For the moment any performer steps on to a stage, they show themselves to be haunted, shot through, with an earthly power – the elemental 'magnetism' – that all bodies possess. To grasp what that power is and to understand how theatre can express it, it is necessary to rethink the meaning of theatricality. In doing so, it becomes possible to see why theatre is a terrestrial medium, an art form that divides bodies from themselves, a materialist space, specially designed for transformation.

Ecologising theatricality

As a consequence of its contradictory usages within a wide variety of disciplines beyond Theatre and Performance Studies, theatricality is often regarded as a problematic term, a word that needs careful definition and contextualisation.[11] Nevertheless, when approached from within the disciplines of Theatre and Performance Studies themselves, there is a general consensus about what theatricality means. At its most basic or subtractive, theatricality, for theatre theorists and practitioners, is generally understood as a mode of showing that makes use of the specific properties of the stage. Features commonly stressed are: the body of the performer, the ephemerality of the event, its location within and organisation of 'real' time and space, and, of course, the fact that theatre is self-consciously constructed for the purposes of being looked at and deciphered by spectators.[12] As Erika Fischer-Lichte notes:

> Theatricality may be defined as a particular mode or using of signs or as a particular kind of semiotic process in which particular signs (human beings and objects of their environment) are employed as signs of signs by their producer or recipients. (Fischer-Lichte, 1995: 88)

In the 1990s, theatricality, for Fischer-Lichte, is conceived as a *wholly* humanist venture, a way of mobilising audiences intellectually, making them critical and self-aware, and thus better able to 'reflect on the conditions underlying and guiding the process by which they construct reality' (Fischer-Lichte, 1995: 104).[13] However, behind this conventionally constructivist or Brechtian view of theatricality as a 'way of seeing', it is also possible to ascertain a potential posthumanist application in Fischer-Lichte's thinking that few have commented on. For as well as reminding us that reality is a semiotic construct, theatricality, as conceived by Fischer-Lichte, discloses the ways in which 'human beings and objects' are haunted by the very thing that disturbed Plato: their *virtual* potential to become other than what they are, to stand in for each other:

> [W]hilst human beings and the objects of their environment in every culture always exist in certain communicative, practical and situative contexts which do not permit a human being to be replaced by another or by an object at random or vice versa, [in theatre by contrast] mobility is the prevailing feature in the case of the human body and the objects from its surroundings when they are used as theatrical signs [*sic*]. Here, a human body can indeed be replaced by another body or even an object, and an object can be replaced by another random object or a human body because in their capacity as theatrical signs, they can signify one another. (Fischer-Lichte, 1995: 88; my addition)

In its capacity to show the workings of the virtual, Fischer-Lichte highlights the sense in which theatricality does not 'ground' the human, but conversely, as Plato also knew, deterritorialises it, revealing its inherent plasticity. In keeping with cybernetic and computational versions of posthumanism, the human, in Fischer-Lichte's understanding of theatricality, is a mere effect of the signifier, the by-product of a binary-focused semiotic system in which everything can be replaced by everything else and where identity is a matter of material context, not metaphysical essence. But as much as this understanding of theatricality certainly troubles the human as a transcendent entity, a being with some supposedly essentialist identity, it does not, however, ecologise it – at least not in the manner that interests me. Not only is language still the dominant paradigm in early Fischer-Lichte (her idea of semiotics is rooted in the idea that signification is a matter of convention, a historically agreed framework of communication) but the human spectator is the agent charged with unpacking signs, a distanced observer standing apart from the perturbing, diffractive play of the virtual which the stage brings so acutely into appearance and which expresses itself through bodies as a force eluding the intelligence.

A more generative insight into the possible ecology of theatricality is offered by the influential work of Josette Féral. Theatricality, for Féral, is not just about staging semiotic transformations, it is also about the affective transmission that occurs when bodies refuse to be reduced to signs: in short, when they insist on their corporeality. Like an electric conductor, it channels a sensate current that implicates spectators physically in a milieu they are both part of and set apart from.[14] A 'pathic exchange' takes place in the theatrical encounter, desire is put into circulation:

> Theatricality can therefore be seen as composed of two different parts: one highlights performance and is made up of the realities of the imaginary; and the other highlights the theatrical and is made up of *specific symbolic structures*. The former originates within the subject and allows his flows of desire to speak; the latter inscribes the subject in the law and in theatrical codes, which is to say, in the symbolic. Theatricality arises from the play between these two realities. From now on it is necessarily a theatricality tied to a desiring subject, a fact which no doubt accounts for the difficulty in defining it. Theatricality cannot *be*, it must *be* for someone. In other words, it is *for the Other*. (Féral, 1997: 297; original italics)

Approached as a 'seam' between the Symbolic and Imaginary, theatricality has no need to stand in for an absent thing, as Fischer-Lichte assumes. In Féral's account, theatricality produces desire by targeting the Imaginary, infecting spectators with an excess that spills over from the proximity of living bodies.[15] This onus on the sensate quality of the Imaginary explains why Féral is so concerned to stress the performative aspects of theatricality, the sense in which the new is made palpable by an affective reordering, a disorientation of the actual. By allowing itself to be framed on a stage, the body is immediately divided from itself and opened out, caught up in an energetic flow that provokes becomings and transformations. Through the syncopated gap or rift that the stage inflicts on the actor's body, a wound is reopened, an ontological chasm disclosed that shows any entity to be 'always more than what it is', available for metamorphosis. Despite my recourse to visual metaphors in the sentence above, the potential contained in this void cannot be seen. It resists the visible because it is the ground on which the visible takes place, a geometry of virtual lines and transversal flows that establishes intensive connections, shattering the boundaries required for the intellectual work of semiotic deconstruction to take place. Theatricality aims to impress not to inform; it is here *and* elsewhere, a doubleness in movement, a stuttering in action. As Féral, in a discussion about the performative dimension of theatricality, explains below:

> For there is nothing to say about performance, nothing to tell yourself, nothing to grasp, project, introject, except for flows, networks, and systems. Everything appears and disappears like a galaxy of 'transitional objects', representing only the failures of representation. To experience performance, one must simultaneously be there and take part in it, while continuing to be an outsider. Performance not only speaks to the mind, but also speaks to the senses [...] and it speaks from subject to subject. It attempts not to tell (like theater), but rather to provoke synaesthetic relationships between subjects. (Féral, 1997: 298)

Féral's perspective on what one could call the 'theatrical performative' has much in common with the ecological rereading of Plato I advanced on pp. 124–7.[16] In both instances, signs are not simply linguistic and neither is the representation of an external reality the primary concern. Theatricality is a disruptor. It touches spectators with the incorporeal power of semblance, the im/material force which, like the magnetism that Plato associated with the actor Ion, exists somewhere between absence and presence, self and other, drawing the spectator in and yet distancing them at the same time. Indeed, in the extent to which the theatrical performative, in Féral's thinking, simultaneously 'cuts together apart' the actual and the virtual, it provides a complex embodiment of the theoretical concepts I introduced in Chapter 3: the nexus that allows Simondon's 'theatre of individuation', Barad's concept of diffraction and Guattari's transversal notion of subjectivity to exist in dialogue with each other. The cogency of this claim is underlined by Féral herself when she defines the actor as 'a subject in process who explores the other he creates', someone who uses the 'displacements' of theatricality 'to put his very self at stake' (Féral, 2002: 99, 100).

Yet, while Féral's understanding of the theatrical performative certainly allows for this non-human aspect of theatricality to emerge, her understanding of otherness remains resolutely within the cultural realm (Féral, 2002: 100). Desire, for her, as indeed for almost all champions of theatricality such as Jonas Barish (1981) and Tracy C. Davis and Thomas Postlewait (2003), is imaginable only as a socio-historical affair, something produced by humans for humans.[17] To get closer to 'a more than human' understanding of theatricality, it is necessary to look elsewhere: namely, to Samuel Weber's *Theatricality as Medium* (2004), a text that surreptitiously ecologises the non-human potential that theatricality always possesses and which both Fischer-Lichte and Féral alighted on without fully unfolding.

For Weber, theatricality fissures space and time. It 'takes place from place', as it were, infusing it with virtuality. As he comments, 'the theatrical entails the intrusion of spatiality within the process of localisation: the fact that the process of being situated has to include (spatial) relationships that it cannot enclose or integrate' (Weber, 2004: 10). By disturbing the reality of place, showing every ground to be perforated with 'a hollow' or 'void' (Weber, 2004: 10), theatricality reminds spectators that the 'stage' is precisely that: a punctuated moment, a suspensive step or passage that derealises and virtualises, highlighting the incompleteness of every object or thing that is placed on it:

> When an event or series of events *takes place* without reducing the place 'taken' to a purely neutral site, then that place reveals itself to be 'a stage', and those events become theatrical happenings. As the gerund here suggests such happenings never take place once and for all but are ongoing [...]. They can be said, then, in a quite literal sense, *to come to pass*. They take place, which means in a particular place, and yet simultaneously also *pass away* – not simply to disappear but to happen somewhere *else*. Out of the dislocations of its repetitions emerges nothing more or less than the *singularity of the theatrical event*. Such theatrical singularity haunts and taunts the western dream of self-identity. (Weber, 2004: 7; original italics)

Weber's final sentence is particularly resonant, especially with respect to the ideas of Barad and Lyotard on spooky entanglements and the im/material, respectively. Through its 'haunting' and 'taunting' of the human subject, its ungrounding of spacetimematter, theatricality exposes what Platonic philosophy is so determined to deny: the preindividual forces of the earth, the neutral gravitational pull that produces de-phasings and becomings, deterritorialising everything that would stand in its path.

In his attempt to convey the dissonance of theatricality, this art form that moves and 'yet goes nowhere' (Weber, 2004: 228), it is significant that Weber should alight on the image of a hand, or more precisely a finger, the digit that has for so long been the figure of/for humanist exceptionalism.[18] Here the hand-finger assemblage is not the organ that would define the human, allowing it to make tools and thus manipulate and mark the earth through the machines it creates. On the contrary, the hand points beyond the human, placing it in contact with some boundless 'outside', a neutral abyss that determines it, unworking its identity according to some unlocatable schema:

> Pointing can be a means of anticipating the seizure and appropriation of what is being pointed at or out, but it can also involve a movement *away* from the familiarity and control of the grasping hand [...]. The 'digital' points away from its immediate manifestation [...]; it signifies something other than what it represents, situated elsewhere. (Weber, 2004: 50–1; original italics)

By showing the performer striving towards something it can never quite coincide with or hold, theatre is evocative of an impossible, non-human intimacy, a medium for the present participle, the tense which best expresses the prodigal expenditure of the earth itself:

> What is curious about the present participle is the way it is both very close and yet irreducibly remote. Since it never adds up to a whole and always remains a part, the participation it entails follows a trajectory [...] that revolves incessantly around a center that is displaced with each turn, never coming full circle, never adding up to a whole, nor even to a simple step forward. (Weber, 2004: 15–16)

Like the earth, the Ur-medium of deterritorialisation, theatricality is founded on a terrestrial temporality that unbalances and dethrones the human. The present participle is a geokinetic: radically open, indeterminate, in the thick of things. And it is ecologically significant, in this context, that Weber should equate the present participle – indeed theatricality in general – with a figure from 'Nature'.

In his analysis of a contemporary staging of 'Autumn River' by the Peking Opera's company, the Liyuan Theatre in 1999, Weber centres on the moment when two humans, Chen and the Boatman, are attempting to negotiate a dangerous channel of water. Following Brecht, who cited a 1935 performance of the same scene in his 1936 essay 'Alienation Effects in Chinese Acting', Weber is struck by the fact that Chen and the Boatman do not approach their task as individuals in the sense that Aristotle specifies in *The Poetics*. They are what Deleuze and Guattari term '*dividuals*', simple hosts for energies and gestures which do not belong to them, but which orientate their movements and motivate their actions:[19]

[T]he performers in this scene do not appear primarily as individuals. Neither passive nor active in the western sense, they demonstrate, quite literally, *a way of being moved* that confounds such oppositions. The skill of the performer allows a movement to be deployed that can never be reduced to the property or product of an individual qua individual. (Weber, 2004: 27; original italics)

In this 'ballet of balance' (Weber, 2004: 28), 'Nature', though only a sign, does not recede for the sake of some uniquely human 'interest story'; rather, it is foregrounded as something to participate in, an actant in a fluid, non-linear world of eddies and tides. Caught in the agency of the river's tow and undertow, the protagonists, Weber proposes, are 'content to respond rather than to impose and resist' (Weber, 2004: 28). In this way, they highlight that their survival depends on surrendering to the greater power of current and wave, with its alternative '*sway of being*', the intrusion of some palpable but indiscernible 'outside' (Weber, 2004: 28; original italics).

This is theatre of a different order – a non-human one. What Weber is touched by and interested in is not the battle for power, prestige and glory that determines the anthropocentric *agon* of Aristotelian theatre, but the ability of bodies to negotiate the movements of an agentic earth. The drama, then, is between human and non-human in 'Autumn River'. The fact that the 'river' is manufactured as an indexical sign, an absence evoked by human gestures alone, is beside the point. The thing to insist on is that the virtual presence of 'river' is acknowledged as the 'shape' that flows through the auditorium, the catalyst for the performance itself, the locus of what Barad names 'response-ability' (Barad, 2010: 251). In this respect, the 'river' is both sign and allegory in Weber's account, a metatheatrical signifier for what theatre is: an apparatus that allows bodies to take off, to depart for somewhere else while always remaining where they are – on an earth that moves, a fissured planet on a sea of magma.

In common with Féral, theatrical vision, for Weber, is always corporeal, and, for that reason, confronted with a non-human enigma, 'a stigma or stain that cannot be cleansed, or otherwise rendered transparent, diaphanous' (Weber, 2004: 7). As a consequence of this 'ineradicable macula', this showing of blindness, theatricality has the potential to make visible an invisibility within the human, short-circuiting the humanist pretensions of narrative drama (Weber, 2004: 7). In lieu of transparency and identification, those aspects of western drama that allow spectators to lose themselves in exceptional others exactly like themselves, the bodies in Weber's reading of Peking Opera appear as opaque, thick substances, always on the verge of 'falling apart', perturbed as they are by the alterity of the earth itself (Weber, 2004: 29). And it is surely noteworthy, from this point of view, that Weber should be so concerned with the moment in 'Autumn River' when Chen's boat almost capsizes. For what Weber discovers in that incident is a theatrical ecology based on 'reciprocity', the sense in which what is at stake for actors and spectators alike is not story but 'rhythm', 'shared but separate movements' that connect people, places and things whilst keeping them separate (Weber, 2004: 29).

The fragility of the theatricalised body is caused by the fact that it is *not* what it appears to be. The actor is a fleshy semblance, a virtuality that disturbs the limits – and so-called reality – of the human that hosts it. Differently from cinema or television that tend to 'bring things closer' to audiences, the 'aura' of the live performer creates an uncanny sense of distance, a suspensive state of (never quite) arriving (Benjamin, 1969: 223). In this vertiginous game of presencing, the live performer plays a permanent game of hide-seek, provoking disjunctive, spasms and tremors that spectators are affected by. The intensity generated by voluminous bodies that ex-hibit themselves, causes reality to flounder, to de-phase; and, in that de-phasing, the pre-individual force that resides within all bodies is experienced by spectators as a dark but vital illumination of the earth's very own 'terrestrial theatre'. In Lyotard's language, the blinded eye that Weber equates with theatre is a prerequisite for and disclosive of the *oikos*, the troublesome stranger that is both within and without, the experience of which deconstructs all oppositions between active and passive, and which leaves one in mid-stream, stuck to the earth. Or as Weber puts it, 'the staging of "Autumn River"' demonstrates how theatre can be the medium of a displacement or dislocation that opens other ways of existing on the planet, 'ways that are not bound to arrive at a final destination – or at least, not too soon. Theatre thus emerges as a powerful medium of the *arrivant*' (Weber, 2004: 29; original italics).

Though he never says so directly, Weber's notion of theatricality is ecological because it positions theatre as a mobile *plateau*, an energised, reciprocal environment where protagonists lack all appetite for oneness, totality and boundedness – those 'values' that capitalist humanism is so compelled to impose on life in order to keep things as they are. In its celebration of divided subjectivity, its acceptance of artifice, theatricality shows life to be positioned somewhere between 'land and water', a flowing 'place' that is earthly because it is always transforming, out of step with itself, indeterminate and incomplete (Weber, 2004: 23).[20] Through his river-soaked and geokinetic concept of theatricality, Weber allows one to grasp that theatre has no need to imitate 'Nature' or to tell ecological stories. Theatre *is* a manifestation of 'Nature'. It plugs into its 'own double', to use Artaud's terminology. It individuates by fissuring spacetimematter. Like the ripple of a current on some 'autumnal river', or in the manner of an earth tremor, theatre produces images, gestures and rhythms that emerge from the opaque surfaces of bodies. These bodies refuse to be fathomed because there is no depth to them, no meaning to discover, no hermeneutic secret to decipher. As an ecological intercessor, theatre is an art of the 'raised curtain', the apparatus that ex-poses what was always already at work within the human from the very start: a constitutive virtuality, a non-humanness, that no one can dispose of, and which, to refer back to Guattari, is so essential to the recreation of our 'mental ecology', the necessary prelude for the arrival of a new planetary people.

As this ecological reading of Weber's analysis of 'Autumn River' underscores, theatricality does more than conjure the presence of an absent river, it gives rise to new universes of ecological value by dislodging the implacable actuality of things as they are, placing reality in movement and gesturing towards an unfixed earth. This is the earth that humans always forget, seduced as they permanently seem to be, by the deceptive presence of the phenomenal, a 'Nature' that appears to be *there* for them, and not, as it actually is, flowing in, through and around them as a kind of im/material wave. Like Chen and the Boatman, spectators are compelled to affirm this theatrical wave, for the affective power contained within it is what allows them to recreate themselves with the earth, not against it.

Conclusion

In this chapter, I have argued for an alternative concept of theatre ecology, one that is forged in the immanence of the theatrical event. When it theatricalises itself, theatre has the capacity to place spectators within a process of individuation, a de-phasing that problematises rigid differences between 'Art' and 'Nature', and truth and appearance. To create a valid theatre ecology is to tap into the earth's own theatrical forces and impulses that can be reached only if the western humanist appropriation of the medium is opened *up* and *out* to an outside that is already inside, approached as a mere fragment of a 'gigantic non-human theatre'.[21] That this idea of theatre ecology is not idle speculation, a figment of an overheated, theoretical imagination, becomes apparent by plotting an ecological course through the history of modern western theatre. As I will show in the next chapter, this history departs from those already charted within Theatre and Performance Studies by insisting on the presence of non-human percepts and affects, those theatrical powers that Artaud calls 'cruelty'.

Notes

1 The UK theatre historian Glynne Wickham's definition of theatre stands as a metonym for a long tradition of western scholarship that typically equates theatre with the human, and, as Hana Worthen points out, with humanism in general (Worthen, 2020: 1–62): '[W]e must constantly remind ourselves that theatre is essentially a social art enhancing and reflecting religious and political beliefs and moral and social concerns as well as literature, music, painting and dance. Indeed, so wide are its terms of reference that theatre has often been used as a metaphor for life itself. As such, theatre is itself a language, coupling verbal with visual images, which assists humanity to understand itself – to define its culture – rather than a craft for the gifted few or a recreation for a privileged elite' (Wickham, 1994: 12). Irrespective of his welcome insistence that theatre cannot be reduced to western drama, nowhere does Wickham suggest or even acknowledge that theatre is also a non-human art form. That is to say, a medium that takes place in time and space; needs the presence of air to transmit words from actor to auditorium; has recourse to objects, props and technology – the 'electric suns' that Brecht spoke of in his cycle *Poems of the Crisis Years 1929–1933* (Brecht, 1976: 176); and, perhaps most obviously, is dependent on the body, that 'queer' assemblage of cells, muscle, and bone that the human subject shares with other creatures, plants and minerals. A similar, exclusively anthropocentric approach to theatre history is found both in Tobin Nelhaus, Bruce McConachie, Carol Fisher Sorgenfrei and Tamara Underinder's third edition of *Theatre Histories: An Introduction* (2016) and in the vast majority of contributions to Christopher Balme and Tracy C. Davis *et al.*'s six-volume study *A Cultural History of Theatre* (Balme and Davis, 2017), whose title is an axiomatic expression of its anthropocentric focus.

2 Martin Heidegger's text '"The Ode on Man" in Sophocles's *Antigone*', published in *Introduction to Metaphysics* (Heidegger, 2014 [1953]), highlights the strangeness inherent in the Chorus's view of 'Man' by translating 'miracle' as 'uncanny' and 'demonic'. While Heidegger's exegesis disrupts the traditional humanist view of 'Man' by stressing its excess and violence, thus potentially endowing the play with a latent environmentalism, it does not, however, question the fact that in Sophocles's play, as indeed in western drama at large, the only creature worthy of attention is the human.

3 Alan Read is a notable exception, as I explained in Chapter 1 (see pp. 47–50). For a useful account of the reluctance of Performance Studies to deal with 'more-than-human performativity', see Chapter 2 of João Florêncio's PhD thesis (Florêncio, 2014: 71–126). Florêncio rightly praises Jon McKenzie for stressing the non-human play of technocratic systems and networks. However, while McKenzie certainly opened performance up to the autopoesis of cybernetic circuits and loops that all life shares, Flôrencio is correct to remind the reader that 'he left untouched the perceived necessity of having humans as the ultimate arbiters of performance' (Florêncio, 2014: 90).

4 In this context, see Alan Read's discussion of Plato's wall as a way of 'evoking, a less metaphysical, more material [...] pre-history of performance' (Read, 2014: 7); and Nicolás Salazar Sutil's critique of Plato's failure to see not only that walls *can* speak, but that language needs walls *to* speak (Sutil, 2018: 111–18).

5 I use dedifferentiation rather than undifferentiation, since to be dedifferentiated is to realise that matter is dynamic, caught up in an endless dialectic between creativity and deconstruction. Undifferentiated matter, by contrast, suggests that only destruction has prevailed and that matter has returned to some amorphous state of self-sameness. I borrow the distinction from Robert Smithson who adopts it from Anton Ehrenzweig's text *The Hidden Order of Art* (Smithson, 1996: 207).

6 There are two moments in Smithson's essay that are particularly resonant. The first is when he talks of how a 'monumental parking lot divided the city in half, turning it into a mirror and reflection – but the mirror kept changing places with the reflection. One never knew what side of the mirror one was on' (Smithson, 1996: 73). The second centres on a children's sand pit, in which 'the irreversibility of eternity' is proved by what he bathetically calls a *jejune* example: a child runs a hundred times in the pit mixing-up the original white sand and turning it grey, and then repeats the experiment but from the opposite direction. As Smithson duly notes, with a kind of anti-Platonic humour, 'the result will not be a restoration of the original division, but a greater degree of greyness and an increase of entropy' (Smithson, 1996: 74). For another work of Smithson's that troubles the relationship between mirror and earth, reflection and matter, see *Yucatan Mirror Displacements 1–9* (1969). In this sited piece, Smithson installed nine mirrors in nine different places in the Yucatan Peninsula in Mexico and insisted that the images did not show the earth as much as come from the earth itself, expressing its fissures and folds.

7 In his writings on the painter Karel Appel, Lyotard notes, again and again, the ways 'in which matter experiments with itself' (Lyotard, 2009a: 101).

8 I am alluding here to Jacques Rancière's critique of Plato's anti-theatricality in *On the Shores of Politics*. Where Plato demands order and oneness, the actor, Rancière proposes, shows that one is 'always out of place' (Rancière, 1995: 11). This spatial anarchy, this doubleness, is precisely why the actor is a figure of politics, for Rancière. Politically, Rancière's argument holds, but it says little about ecology. To ecologise his position, one needs to stress the role of matter and energy, the play of the pre-individual force that Simondon talks of.

9 In his reading of Edgar Allan Poe's eponymously titled short story, Lacan explains how the 'purloined letter' is like the psychoanalytical symptom: it hides in plain sight (Lacan, 1972: 39–72).

10 Interestingly, Guattari also uses a geological image to highlight the non-humanness that courses through the 'interior', psychical space of the subject. According to Guattari, the subject's 'Territories of Existence [...] drift in relation to each other like tectonic plates under continents' (Guattari, 2000: 24).

11 A good sense of the troubling ambiguity of the word, its dizzying capacity to contradict itself, is provided by Tracy C. Davis and Thomas Postlewait in their excellent and comprehensive collection *Theatricality* (2003): '[T]he idea of

Concept   139

theatricality has achieved an extraordinary range of meanings, making it everything from an act to an attitude, a style to a semiotic system, a medium to a message. It is a sign empty of meaning; it is the meaning of all signs. Depending upon one's perspective, it can be dismissed as little more than a self-referential gesture or it can be embraced as a definitive feature of communication' (Davis and Postlewait, 2003: 1).

12 This generic, largely non-controversial understanding is prevalent in supporters of theatricality, such as Roland Barthes (1964), Elizabeth Burns (1972), Jonas Barish (1981), Jean Alter (1990), Marvin Carlson (2002), Andrew Quick and Richard Rushton (2019) as well as detractors like J. L. Austin (1962), Michael Fried (1967), and Judith Butler (1988). In these readings, theatricality is a property of theatre and pertains to the act of constructing a spatio-temporal event that draws attention to itself in the very act of signifying.

13 On the surface, Fischer-Lichte's work in the 2000s is less anthropocentric in focus, concentrated, as it is, on unpacking the performative significance of ecological terms such as autopoesis and feedback loops. Indeed, *The Transformative Power of Performance: A New Aesthetics* has much to offer ecocritics, even though it does not deal with ecology in any kind of frontal way, until its perfunctory and elliptical final pages. Differently from her earlier work, Fischer-Lichte is concerned in this text with affects and atmospheres, and her use of terminology from second-wave systems theory provides an excellent, aesthetic addendum to how Kershaw uses cybernetics. In Fischer-Lichte's account, performance is a materialist and affective artform that transforms spectators by approaching them 'as living organism[s], constantly engaged in a process of becoming' (Fischer-Lichte, 2008: 92). The problem, however, with Fischer-Lichte's argument from an ecological perspective is that she holds on to a proper view of the human. The subject may well be engaged in a process of becoming through performance, but it is always becoming itself. Despite its ecological vocabulary, performance, then, is once again posited in Fischer-Lichte's text as an art of self-presence, a medium where bodies can be reconciled to minds, and where self and other might exist in a relationship of enchanted *communitas*. In its search for oneness and propriety, the new aesthetics of performance that Fischer-Lichte so lucidly expounds remains tied, ultimately, to a conventional humanist ontology and epistemology.

14 Jean Alter's notion of the 'performant function' of theatre, which, in his 'binocular view', exists alongside its referential or symbolic side, has much in common with Féral's view of theatricality. In both instances, theatricality is founded not simply on a symbolic lack but on the expression of something positive, a transmission of energy, an affective catalyst, some unspeakable quiddity (Alter, 1990: 32).

15 Although Féral 'recuperates' theatricality, she nevertheless maintains a critical attitude towards theatre, preferring the greater immediacy of performance. In that respect, and somewhat ironically, she comes close, at times, to investing in the 'anti-theatrical prejudice' shown by Michael Fried in the 1967 essay 'Art and Objecthood'. The difference being, of course, is that she has no concern with

'grace', a type of presence that, as Fried has it, allows the spectator to transcend space and time and to exist as unitary substance (Fried, 1967: 23).

16 Left unqualified, Féral's early notion of theatricality can be confusing, since it is a property of performance without ever being equivalent to it. Where performance, for Féral, completes itself in an act of immediate self-presence, theatricality always leaves a remainder and demands repetition. But, as I have been arguing, theatricality's deferred economy does not mean that it is any less sensate than performance. It is simply that its affective economy is predicated on a different logic, one that insists on the ungraspable movement of the virtual, an im/material 'touch' that can never be held or communed with in any kind of absolute sense. Theatricality undermines performance as an art of immediacy, something that would restore presence, in the philosophical sense of the word.

17 Worthen is excellent on explaining how Barish's 'anti anti-theatrical prejucide' in his landmark text of the same name is bound up with a Cold War commitment to liberal humanist values, the production of a democratic ensemble that needs theatricality in order to realise itself, to see itself externalised and represented as other (Worthen, 2020: 26–31).

18 In contrast, Haraway's posthuman notion of storytelling depends on the hand, the organ that holds together the string figures in her account of what she calls 'SF', in an attempt to bring together string-figuring with science-fiction and speculative feminism, amongst other things (Haraway, 2016: 10).

19 In 'Postscript on Control Societies', Deleuze's notion of the dividual refers, negatively, to the process whereby individuals are seen as quantified data sets, mere information. There is, however also a more emancipatory aspect to dividuation, where the dividual is able to escape the fixed territory in which identity is contained (Deleuze, 1995: 180–2).

20 Jean Baudrillard provides an apt description of how perception stumbles in its attempt to capture the earth, and, by extension, theatricality, when he says that human perception is only able to conceive of the earth as 'a mental object', a distortion of an actuality that 'cannot be encountered' (Baudrillard, 1996: 1).

21 I adapt this term from the title of a collection by Alice Oswald and Paul Keegan on 'weather writing', who, in turn, borrow it from Virginia Woolf's thoughts on the movement of clouds (Woolf in Oswald and Keegan, 2020: 5).

5

History (Artaud's cruel ecology)

Introduction

As I intimated towards the end of Chapter 4, there is no requirement to invent a new paradigm of ecological practice for Theatre and Performance Studies; the more judicious move is to perform an act of excavation, revealing the ecological potential that modern western drama has long been in possession of. In order to chart that potential, I divide the argument of the chapter into two movements. First, I explicate the 'more than human' genealogy of western theatre advanced by Michel Corvin in his undervalued text *L'Homme en trop: l'abhumanisme dans le théâtre contemporain* (2014). Second, I concentrate on the practitioner-theorist who has arguably done more than anyone to create an immanent ecology for contemporary theatre: Antonin Artaud. Because Artaud has proved such an influential and, at times, divisive figure, it is imperative not only to ecologise his concepts but to take issue with landmark readings of his work both within and beyond Theatre and Performance Studies.[1] This need for rigour and explanation accounts for the circuitous shape of this chapter, caused by the need to defend Artaud's cruel ecology from critics who may well be reluctant to acknowledge, let alone, endorse it.[2]

Abhumanism: *L'Homme en trop*

Despite showing no interest in environmentalism, Corvin's *L'Homme en trop: l'abhumanisme dans le théâtre contemporain* offers key insights into theatre's immanent ecology.[3] Its purchase resides in how it tracks the theatrical legacy of Symbolism, which, as Corvin has it, was the first avant-garde art movement to break with the human-centred dramaturgies of the western tradition. Differently from Classical and Shakespearean tragedy (and later naturalism and realism), both of which 'made an exalted praising of man' ('faire un éloge exalté de l'homme') (2014: 13; my translation), Symbolism, with its interest in androids, silences and atmospheres, initiated a new practice of theatre; one that Corvin refers to as abhumanism, after the poet and playwright Jacques Audiberti, the inventor of the term.[4]

In Corvin's dramatic recalibration of Audiberti's original thesis, abhumanism is no longer a philosophical concept, an attack on the principles of Enlightenment thought. Above all else, it is a mode of dramaturgy, an anti-representational attempt to explode the humanism inherent in Aristotelian ways of depicting the human subject on stage.[5] As Corvin has it, abhumanism departs from humanist theatre in two ways. First, it insists on the quality of theatre that Aristotelian mimesis does its utmost to foreclose: the disruptive power of theatricality.[6] Second, it attempts to populate the stage with a whole host of non-human others, to signify a more democratic and diverse world – one in which the category of the human itself is queered, expanded, remade. Abhumanism, Corvin tells us, is present in the puppets and masks of Jarry and Craig; the costumes and geometries of Dada and Bauhaus; the dreamscapes of Surrealism; the gestures and cries of Artaud; the distorted bodies of Beckett; the hieroglyphs or living symbols of Genet; the dramatic deconstructions of Martin Crimp; the sexualised cyborgs and grotesques of Gisèle Vienne; the voids and silences of Valère Novarina's and Claude Régy's actors; and the animals of Socìetas Rafaello Sanzio.

Irrespective of the grammatical stress placed on the Latinate 'ab', a prefix habitually associated with the act of abandoning, reducing or 'worsening', abhumanism does not just signify 'lessness', 'inferiority' or 'poverty', in Corvin's definition; and neither does it correlate with the understanding of posthumanism I discussed in Chapter 2. Abhumanism's specificity, for Corvin, is located in the excessiveness of the human, in the power the actor has to become almost anything – the *en trop* ('too muchness') signalled in the title of his text. In its theatrical incarnation, the abhuman subject is a being that strays, finding itself mixed up with animals, gods, machines, ghosts, linguistic systems, cyborgs, puppets, geometries, surfaces, objects and corpses. However, a deep ambivalence runs through abhumanist theatre like an unravelling thread. For as much as it troubles dominant definitions of what a human being is, showing it to be a mere thing amidst a plethora of other things, it also endows the human with great power. As Corvin puts it in a long list of prefixes, everything in abhumanist theatre speaks of a human subject whose only essence is to have no essence, a plastic creature, an animal that is fated to never be itself.[7]

As with Weber's writing on theatricality, nothing explicitly ecological is articulated by Corvin's Symbolist-inspired history of abhumanism. His interest is not in the history of the earth, but in the history of dramaturgical experimentation, in tracing the different strategies used by modern western theatre makers, from roughly the 1880s onwards. Yet there is nevertheless something in abhumanism as a stylistics, if not a philosophy, that has crucial ecological relevance.[8] For if theatre, as I pointed out in Chapter 4, is an art form that stages, more palpably than any other, the theatrical quality of ecology, the way in which life is a 'drama of individuation', constantly varying and dissenting from itself, then abhumanist dramaturgies have the potential to make that process tangible, to bring it to the surface, and, in some cases, to act as agents on its behalf. In consequence, Corvin's important but indirect contribution to theatre ecology is to have pointed to a history of contemporary theatre whose ecological affordances are already there: notably, when it dares to make a virtue of theatricality and to point to a more expansive non-human cosmos, a world of creatures and things in permanent transformation. In abhumanism, theatre is no longer a place for psychologising, explanation or storytelling, it is a site for troubling the boundaries of the human, a desire to include other beings and actants on the stage.[9]

At this point, though, a problem arises in Corvin's thinking that explains why I do not simply adopt his terminology wholesale. For as much as Corvin's study proposes, in embryonic form, a way of conceiving an alternative practice of theatre ecology rooted in a decidedly theatrical aesthetic, it is compromised by the fact that many of the versions of abhumanism he points to appear to be decidedly humanist in remit and orientation. Corvin's 'mistake' is foundational to his argument. The starting point of his thesis is based on a concept of dramatic humanism that he finds in Aristotle's *Poetics*. Humanism, here, is predicated on verisimilitude: the idea that the character should resemble its living model as closely as possible and represent the human being as a self-coincident creature, able to think, act and judge on its own autonomous terms. However, recalling Colebrook's critique of posthumanist ecology in Chapter 2 (see pp. 81–3), humanism is not just about mimesis, in which humans stand in for other humans, as Corvin contends. More accurately and expansively, it is a credo in which the human looks for oneness, both within itself and with everything else in the universe. In other words, and to put it succinctly, humanism is a worldview where transcendence, plenitude and identity reign. Although it is conventionally associated with the human subject, humanism, bizarrely, does not need 'flesh and blood' human representatives for its logic to pertain. Its fundamental requirement is for a consciousness that is able to recognise itself in whatever form it takes. Perversely, to bring those ontotheological values into the human realm, no human actor is required. Anything will do: an animal, a robot, a puppet, a god, a rock.[10] The prerequisite is that human consciousness is able to project itself on to its non-human other and lose itself within it. In this anthropomorphic operation, theatrical notions of 'as if' are all important. The non-human other does not exist in its own terms, in its opacity. Rather, it exists as 'if it were like a human', serving as a supplement to the human, permitting consciousness to overcome anything that challenges its unity, a projective move that is exemplified, theatrically, by Edward Gordon Craig's theory of the Über-Marionette – a text that is central to Corvin's argument.

In the influential 1908 essay 'The Actor and the Über-Marionette', Craig explains that he is attracted to the puppet because it is devoid of the theatrical oscillation, the pre-individual flow that disrupts the self-presence of the human performer. The puppet's beauty, its capacity to produce the kind of communion that Craig is so seduced by, derives from its possession of an essence that is immutable – the fact that it is 'dead', in other words.[11] It is precisely this 'deadness' that appeals to Craig. Deadliness is the quality that endows the puppet with the aura of permanence and wholeness, a transcendent consciousness that spectators can project their own consciousness on to. Curiously, then, the puppet, through its deadness, shows itself to be the perfect human – at least in Colebrook's critique of that term and credo. Because it is an embodiment of being as opposed to becoming, the puppet possesses what humanism desires: a self-contained and holistic identity from which all difference has been expunged.

Further light is shed on the strange humanism of the puppet by Heinrich von Kleist at the end of his beautiful 1810 essay 'On the Marionette Theatre'. According to Kleist, transcendence – or what he calls 'grace' – is not achieved in theatre by humans identifying with 'fallen' creatures like themselves; it is brought about through the staging of 'bodily structures' that are devoid of self-conscious: namely, animals or, better still, 'mechanical puppets' (Kleist, 1972: 26). As Kleist proposes, the lack of self-consciousness in animals and puppets is not, in any way, a flaw. Conversely, if amazingly, it is the thing that allows them to be a superconsciousness, a sort of hollow or perfect mirror that permits the human spectators to overcome their alienation, to be whole.[12] In their calls for an inorganic, puppet theatre, Kleist and Craig problematise Corvin's theory of abhumanist theatre. Ironically, they show that humanism is not at all troubled by replacing human actors with non-human actants. On the contrary, a marionette theatre is the site where humanism reaches its apogee. Primarily because it replaces division with oneness, flesh with machine, finitude with infinitude, and life with death. Where the human actor shows its artifice and thus alerts us to its doubleness, to the fact that it is not what it appears to be, the marionette by insisting on artifice shows itself to be exactly what it is. In it, consciousness and being are identical. Transparency defeats opacity; oneness is all.

As this intimacy with death reveals, there is an implicit violence at work in humanism's quest for oneness. Anything that stands in the way of the humanist search for transcendence is immediately exterminated. If consciousness cannot control the earth, then the earth must be sacrificed. Such aggressiveness has important implications for the ecological and political stake of abhumanism. The last thing one wants from an ecotheatrical perspective is to use Corvin's abhumanist ideas to fulfil a reactionary agenda. This would be one in which the extreme plasticity of the human is the thing that constitutes its freedom, the foundation of its being. The danger here, of course, is that plasticity, as Corvin has it, is simply another word for human exceptionalism, the negative quality that underpins its protean power to become everything, to colonise the planet.

The ethical problem in Corvin's abhumanist theatre is borne out by Lyotard, when the latter talks of two types of 'inhuman' that ought never to be confused. On the one hand, the 'inhumanity of the system' that exploits the 'absence of any defining property' specific to the human to institute a stultifying world of competition, development and managerialism; and, on the other, 'the infinitely secret' inhumanity, which is associated with 'the anguish of a mind haunted by a familiar and 'unknown guest' – an *oikos* – that 'thinks', 'feels' and utters 'promises of the possible' (Lyotard, 1991: 2–4). Crucially, the anguish of this 'secret inhumanity' is the emotion that commits humans to terrestrial life, allowing for the intensity of feeling to take precedence over the dreary consolation of knowledge. In Lyotard, as with Colebrook, the human's dependence on the earth does not erase its difference. It holds it in tension as a disjunctive unity, an open totality without synthesis. Humans have never been anything other than posthuman, for Lyotard and Colebrook, creatures that constantly fail to become one with themselves. So, while some of Corvin's prime examples of abhumanism – Symbolism's androids, Craig's puppets – may well depart from dramaturgical humanism in mimetic terms by embracing the inevitable theatricality of theatre, they do little to challenge the ontological and epistemological assumptions of philosophical humanism – the belief that the human is somehow able to transcend its own alterity and thus know itself by finding the truth of consciousness reflected in a person or object that is purportedly other than itself.

For the theatre ecologist, Corvin's abhumanist genealogy of western theatre is ambiguous, fundamentally dissonant. As much as it offers the potential for thinking through what the poetics of an ecological stage might look like, his notion of what constitutes the abhuman is lacking in ecological ethics and politics. Too often in his thinking, the human remains stubbornly central. Not a creature of weakness, but a creature of power, a Promethean-like being. Consequently, instead of looking at how humans confuse themselves with a plethora of non-human others, it is more useful from an ecological perspective, to investigate a possibility that Corvin downplays: that is, the way in which some forms of contemporary western performance highlight the non-human that is already within the human, the ungraspable limit that was always already there from the beginning. The shift, in other words, needs to be from an abhumanist stage to a posthumanist one, a departure that accords heightened importance to the work of Artaud.

Differently from Corvin, who is content to fold Artaud into a general discussion about the development of abhumanist aesthetics, I prefer to stress something more specific: Artaud's love of the earth, his desire to show human beings troubled by non-human forces which they are unable to control or identify with.[13] In an ecological *détournement* of Plato, Artaud figures theatre as a terrestrial assemblage, an affair of bodies, rhythms and affects – something telluric, cosmic, more than consciousness. In Artaud's thought, theatre is the site where traditional borders separating humans from animals come apart and where discrete anatomies are shattered. Evidence for that proclamation is found in the celebrated analogy that Artaud draws between theatre spectating and snake charming:

> Snakes do not react to music because of the mental ideas it produces in them, but because they are long, they lie coiled on the ground and their bodies are in contact with the ground along almost their entire length. And the musical vibrations communicated to the ground affect them as a very subtle, very long message. Well, I propose to treat the audience just like those charmed snakes and to bring them back to the subtlest ideas through their anatomies. (Artaud, 1958: 81)

Anticipating Lyotard's view of the *oikos*, Artaud realises that the anatomy of spectators cannot be remade without pain and suffering. To embrace a reptilian attachment to the earth is to live exposed, to affirm what Artaud calls 'cruelty': 'an appetite for live, a cosmic rigour and implacable necessity [...] from whose ineluctable necessity life could not continue. [...] Everything that acts is cruel' (Artaud, 1958: 102, 85). Irrespective of what some critics have claimed, Artaudian cruelty is not opposed to a progressive ethics and politics. It is advanced in the name of a different politics, one that escapes the image of a 'sustainable future' promoted by almost all forms of mainstream environmentalism, a spurious type of global custodianship that Gayatri Chakravorty Spivak has perceptively critiqued as 'feudality without feudalism' (Spivak, 2014: 1223).

Artaud's ecology, like Spivak's, has no interest 'in keeping geology safe for good imperialism' (Spivak, 2014: 1223). Rather, it is committed to the production of a new people, whose planetary being-in-common transcends any narrow identification with nation, class or ideology. In this heterogenous ecology, the thing that matters is to encounter an existential disjunction that conflates suffering with creativity, an individuation-process that demands severance from both humanist consolation *and* abhumanist celebration. For, as Guattari reminds us, 'rupture is at the heart of all ecological praxes', the 'catalyst' for a different mental ecology (Guattari, 2000: 30, 45).[14] The rupture evoked by Guattari is non-human, the passionate presence of a pre-individual force that runs through and across all bodies, including human ones. While Corvin is sometimes attuned to this force in *L'Homme en trop* (I think here of his readings of Beckett and Genet), more often than not, his concept of abhumanism obscures it, concerned, as he is, with tracing the human's representational excess, its capacity to transcend itself through the projection of its consciousness onto a non-human body that stands beyond it.

Artaud is a posthumanist not an abhumanist.[15] He knows that the outside is already inside as a 'double' or 'shadow' that refuses to disappear, giving rise to a subject whose desire for transcendence is constantly thwarted and that no puppet or machine can assuage nor mask. Where Corvin sets no limit to human ingenuity, revelling in the anthropomorphic capacity of the human to surpass itself, Artaud knows that there is something in the human that places limits on its freedom. The human is undone by the non-human that constitutes it and sets it reeling, a doubleness that it can never catch up with or hold. The drawback in Corvin's abhumanist theory of theatre is to have neglected this internal opacity, to have equated the non-human with an external figure, something projected by the human beyond itself as an anthropomorphic object or shape that it can colonise vampirically, while all the time remaining integral and self-present. In abhumanist theatre, the *anthropos* controls the world by reducing everything else in it to the measure of the human imagination. The imaginative power of the human is the very thing that Artaud attacks, and it is this willingness to inflict cruelty that makes him into a theatre ecologist.

Cruelty

Cruelty is a constant presence throughout Artaud's diverse body of work.[16] It starts with his experiments in theatre in the 1920s and 1930s, continues in his drawings and writings at Rodez in the 1940s,[17] and ends with his final performance *Pour en finir avec le jugement de Dieu* (*To Have Done with the Judgement of God*), a radio piece that was commissioned by Fernand Pouey in 1947 and then promptly banned by the station controller Vladimir Porché.[18] Depressed by the censorship imposed by Radio Diffusion Française, Artaud, in one of his last letters to his friend and collaborator Paule Thévenin, defines cruelty as the 'very *genesis* of creation', a 'theatre of blood', in which 'actors are not performing they are doing' (Artaud, 1976: 584–5). Crucially, 'blood' is not spilled by cutting the flesh, as one finds in the practices of so many contemporary performance practitioners and body artists who are too quick to establish (mistaken) allegiances with Artaud. It is reached by im/material means, through the deliberate construction of artificial dramaturgies that rigorously hollow themselves out, communicating their own virtual excess, touching bodies with cosmic forces. Neither skin nor blood is a solely physical entity, in Artaud's aesthetic. They are portals, corporeal zones for connecting with larger, indiscernible flows, liberating the flux of the pre-individual, 'that fragile fluctuating centre which forms never reach' (Artaud, 1958: 13). To set that inaccessible 'centre' free, cruel theatre demands deliberate and careful organisation. Nothing is left to chance. 'Everything', Artaud demands in *To Have Done with the Judgement of God*, 'must be arranged to a hair in a fulminating order' (Artaud, 1976: 555).

This explanation leaves one in a better position to understand what Artaud means when he talks of theatre as 'a doing', a cruel praxis. To attend the theatre, for Artaud, is not to project one's humanness onto others; it is to participate in a becoming that is also a return, to feel one's body disfigured, as in the paintings of Van Gogh, by the 'affirmation of a terrible and otherwise inescapable necessity' (Artaud, 1958: 102) – a non-human force that has always been there, a suffering it is unable to deny. Theatre's capacity to host and disseminate cruelty explains why Artaud is able to say, in complete seriousness, that the actor hosts 'the whole of nature' (Artaud, 1958: 86). By which he means not just organic objects or images that one can represent and gaze at, but, more fundamentally, invisible energies and im/material sensation that the actor is implicated in.[19] By expressing 'external forces' that possess them (Artaud, 1958: 82), performers show themselves, like Plato's actor Ion in Chapter 4 (see pp. 126–7), to be already part of a larger, expanded theatre ecology, channelling elemental frequencies and vibrations that oscillate, pharmacologically, between creativity and destruction, and which divide bodies from themselves: 'Everything that acts is cruelty. It is upon this idea of extreme action, pushed beyond all limits, that theatre must be built' (Artaud, 1958: 85).

Opposed to those contemporary practitioners and theorists who look to equate ecotheatre with messages of environmental care, 'Nature' is not a benign presence on Artaud's mobile *plateau*, something to fall in love with or be enchanted by; it is an excess or burden that inflicts violence and distress: 'If all the attempts of Artaud bring us back to theatre', Alan Virmaux explains, 'it is because theatre shows the fundamental distress of Nature' (in Visser, 2021: 106). Artaudian 'Nature' has little in common with Romantic notions of communion, stewardship and propriety.

Like Latour's monstrous rereading of James Lovelock's notion of Gaia,[20] 'Nature' is unpredictable, violent and irreparable, for Artaud.[21] To live with it, one is compelled to accept its agency and anarchy, to respond to it as an 'unnatural' force, an energy that creates on its own terms. A way of thinking which explains why Artaud finds no contradiction in comparing the effects of theatre to those of volcanoes, earthquakes, the 'crushings and grindings' of plate tectonics (Artaud, 1958: 103).

Like a terrestrial artist, Artaud's earth does not know what it can do, a planet convulsed, situated 'at the incandescent edges of the future' (Artaud, 1958: 51). On it, there is no hope of transcendence or plenitude for the *anthropos*. 'Nature', he says, is in 'perpetual conflict, a spasm in which life is continuously lacerated' (Artaud, 1958: 92). Such a cruel ecology is far from reassuring. It demands a radical recalibration of the human's place in 'Nature', especially at a moment when ecomodernist and technocratic versions of the Anthropocene have supposedly eradicated 'Nature's' power completely, erroneously situating the human as the only actant with power, a geolithic agent able to terraform the earth and potentially other planets, too. But to gain some sense of the progressiveness of Artaud's ecology – one in which the human is decentred but never erased nor negated – a series of preliminary steps need to be undertaken. By proceeding with patience, it becomes possible to rethink the logic of influential interpretations that have seen nothing recuperative in Artaud's cruelty, accusing him only of nihilism, Fascism and colonial appropriation.

In defence of cruelty: against capitalist sorcery

In Jane Goodall's 1994 monograph *Artaud and the Gnostic Drama*, Artaud's theatre is read as an exercise in reverse purification, an obscure, theological attempt to restore the universe to its proper state of unsullied, non-being in which God communes silently and auto-affectionally with himself.[22] As Goodall has it, Artaud's Gnosticism is predicated on the belief 'that an estranged or alien self resides in each human being as a spark of the dispersed pneuma waiting to be released from the corporeal form that prevents it from being reunited with the first great Life, the hidden God' (Goodall, 1994: 15). This hidden God is far from being the molecular God of terrestrial becomings that I have suggested is at play in Artaud's affirmation of the earth. It is radically transcendent, dwelling in the 'Pleroma', a spiritual space located 'in the outer reaches of the cosmos', beyond 'the mundane prison' of the earth (Goodall, 1994: 15). As a point of passage, a 'crossover' point, theatre is imagined by Artaud, Goodall contends, as a site of dialectical negation, a vehicle for leaving the divided, corporeal human subject behind, for escaping the flesh and exiting the base materiality of the earth. In Goodall's reading, cruelty is in no way a prelude for ecology; it is a negative theology, a Gnostic technique for communion with a sacred being, the rediscovery of unity (Goodall, 1994: 15).

Though equally dubious of its terrestrial benefits, Kimberly Jannarone's critique of Artaud's cruelty is more historically and politically grounded. As she sees it, and expanding on an earlier critical intervention by Naomi Greene (1994), Artaud's desire to inflict cruelty on spectators is indicative of a dark, vitalist '*mal du siècle*', a fascistic propulsion to invest in an anonymous, irrational life principle at the expense of anything considered weak or degenerate (Jannarone, 2010: 23):[23]

> Artaud's images have been read as metaphorical, taking their power from ahistorical poetic license. Yet the specificity of Artaud's catalogues of catastrophe is not entirely metaphorical or out of history. His rhetoric and imagery mirror the ravages of World War I and the thinking of those who called for more, who, within a few years, coalesced their irrationalism and mysticism into a body of ideas that fed directly into fascism. (Jannarone, 2010: 23)

Jannarone is right to say that Artaud's cruelty is more than metaphorical. She is also correct when she hones in on his vitalist critique of reason and volition but she is mistaken about its historical relationship to Fascism, no matter how aggressive Artaud's attack on what he calls 'digestive humanity' may sometimes appear (Artaud, 1976: 519). Not every attack on reason is necessarily Fascistic, and neither is every contagion negative.[24] One thinks, for instance, of the composting ecologies of Donna Haraway and Lynn Margulis, for whom contamination is a necessary prerequisite for life itself,[25] and also of Artaud's own outraged critique of the Cold War military-industrial complexes of the US and USSR in *To Have Done with the Judgement of God*:

> And war is wonderful, isn't it? For it's war, isn't it, that the Americans have been preparing for and are preparing for this way step by step. In order to defend this senseless manufacture from all competition that could not fail to arise on all sides, one must have soldiers, armies, airplanes, battleships, hence this sperm which it seems the governments of America have had the effrontery to think of. For we have more than one enemy lying in wait for us, my son, we, the born capitalists, and among these enemies Stalin's Russia which also doesn't lack armed men. (Artaud, 1976: 556)

Jannarone's one-sidedness spills over too, albeit with more justification this time, into Artaud's attraction to non-European cultural practices, which she regards as 'a classic example of orientalism', an attempt to halt the decline of the West by pillaging different cultural traditions and belief systems (Jannarone, 2010: 59).[26] For Jannarone, Artaud 'fetishizes the other', precisely 'because he does not – and does not attempt to – understand them' (Jannarone, 2010: 59). Like a reactionary ethnographer from the nineteenth century or new age guru from the twenty-first, Artaud is attracted, in Jannarone's account, to Balinese dance, the peyote rituals of Tarahumara Indians in northern Mexico and the elaborate death rites of the Tibetans solely for his own ends. They offered, she avers, 'a way out of history, a path to "Myth" – with a capital M, something beyond embodiment, beyond time and place' (Jannarone, 2010: 59).

Although there is little doubt that Artaud's thought is orientalising – it is hard to deny that he projects his own desires and fantasises on to Balinese performers and Tarahumara Indians in order to define them – to stop at this judgement, as Jannarone does, is to forget the complex tension in his thinking, a generative dissonance that James Clifford in his pivotal, 1988 collection *The Predicament of Culture: Twentieth-Century Ethnography, Literature, and Art* names as 'ethnographic surrealism' (Clifford, 1988: 117).[27] This is a way of seeing and perceiving that, while it problematically looks to 'primitive societies' for new possibilities of being, does so in a manner that is not only very different from earlier ethnographic models, but that self-consciously places western imperialism in question in the process. By 'making the familiar strange', ethnographic surrealism fissures the solidity of the colonialist ground it occupies, opening the spectator to an alterity that it had initially projected on to the Indigenous other. Clifford explains:

> Unlike the exoticism of the nineteenth century, which departed from a more or less confident cultural order in search of a temporary frisson, a circumscribed experience of the bizarre, modern surrealism and ethnography began with a reality deeply in question. Others appeared now as serious human alternatives; modern cultural relativism became possible [...]. The 'primitive' societies of the present were increasingly available as aesthetic, cosmological, and scientific resources [...]. Below (psychologically) and beyond (geographically) ordinary reality, there existed an other reality. (Clifford, 1988: 120)

Clifford's comments allow another story to be told about Artaud, especially when his excoriation of US biopolitics and colonialism in *To Have Done with the Judgement of God* is taken into account, as I do on pp. 156–7 in this chapter.[28] This is a story that places Artaud within a larger political and ecological history, a narrative where the colonisation of the Americas in the fifteenth and sixteenth centuries – what the geographers Simon Lewis and Mark Maslin term the 'Colombian Exchange' (2015) – anticipates and meets up with the environmental degradation of the twentieth century, the chemical century that, as Peter Sloterdijk has shown, started in earnest with the release of mustard gas in the trenches at Ypres in April 1915 and ended, globally, in the polluted landscapes of the so-called Anthropocene (Sloterdijk, 2009: 9).[29]

In contrast to Jannarone whose reading of Artaud's 'primitivism' is ethically cogent but politically ossifying, I prefer to stress its emancipatory potential. For all its problems, Artaud's aim in opposing the ritual practices of the Tarahumara Indians to capitalism's planetary war was never just personal, the individual ravings of a mad man on the lookout for some solitary redemption through the exoticised other.[30] Rather, as I will presently show, the stated aim was to instigate a break with what Artaud prophetically named the 'microbial noxiousness' of western ontotheology (Artaud, 1992: 327). The intention being to allow for a more intimate, less exploitative relationship with human and non-human 'Nature', in which all of planetary life is 'liberated' and where 'man' (*sic*) is only a 'reflection' of a greater process of terrestrial becoming (Artaud, 1958: 116).

Irrespective of its reifying utterances, it is inaccurate to claim that Artaud's 'Orientalism' is motivated by primitivist nostalgia for 'savage cultures', at the convenient moment in which these cultures are being lost, as Claude Lévi-Strauss expressed in *Triste tropiques* (1955). A closer reading of his texts shows that the ambition is to effect a radical reappraisal of capitalism's project in the here and now; to demonstrate, in the most heretical and empirical ways possible, that the metaphysical nexus upon which capitalist, humanist subjectivity is predicated – progress, universality, reason – contains an insidious and inevitable colonialism, a will to enslave and pollute that can only result in a planetary conflagration. Significantly, in the 1936 essay 'Theatre and Culture', Artaud saw this conflagration as being characterised by 'the revenge of things' (Artaud, 1958: 9).

In parallel with contemporary thinkers such as Julie Cruickshank (2005), James Clifford, Isabelle Stengers, Déborah Danowski and Eduardo Viveiros de Castro, and José Gil (1998), there is much in Artaud that seeks to 'take indigeneity seriously', finding ways of thinking in tandem with it, affirming its potential for a new planetary commons (Gil, 1998: xi).[31] It is extremely significant, in this context, that the support Artaud gives to the Mexican Revolution in the 1930s is contingent on its willingness to embrace Mexico's Indigenous populations whose cultures and practices, at that time, were largely marginalised by artists and progressive thinkers in the drive to modernise the country (Artaud, 1976: 372). In the text 'What I Came to Mexico to Do', Artaud emphasises, again and again, that Mexico has something 'enormous to teach the world' (Artaud, 1976: 373), but only if it manages to pay greater attention to its 'indigenous soul' (Artaud, 1976: 369). The teaching that Artaud is referring to is both cosmopolitical and ecological, a lesson that rejects the binary thinking of western modernity whose 'superstition about progress', Artaud insists, has resulted in an irrational drive to impose its own discourse as the only discourse (Artaud, 1976: 370).

Artaud is not just interested, as an orientalist would be, in denying the Indigenous peoples of Mexico a history, in consigning them to some Edenic essence (the 'noble savage'). His concern is with diversifying technologies, in offering alternative ways of being on the earth. Like Walter Benjamin's notion of history, the impetus behind Artaud's affinity with Indigenous ways of being in Mexico, as it was in his trip to Ireland in 1938, was to 'blast the past into the future', discovering virtual possibilities that remained latent in the present in order to recreate the world anew, to trouble linear myths of progress that have for so long placed the western subject at the apex of history:

> In short, it is a question of reviving the old sacred idea, the great idea of pagan pantheism, this time in a form that will no longer be religious but scientific. True pantheism is not a philosophical system, it is merely a *dynamic investigation* of the universe. (Artaud, 1976: 373; original italics)

Artaud fleshes the same idea out in To Have Done with the Judgement of God when he says that 'confronted with ersatz people' who industrially farm their own animals for food, 'I prefer the people who eat right out of the earth that gives birth to them. I am speaking of the Tarahumaras who eat peyote [...] and kill the sun in order to establish the kingdom of black night' (Artaud, 1992: 313). It is important to grasp what Artaud is saying, here, to contextualise his images and metaphors. For Artaud is not just engaging in a normative, neo-colonial binary opposing 'Nature' to 'Culture', the organic to the inorganic, authentic to inauthentic. He is, *pace* Clifford's ethnographic surrealism, bringing into conflict two different modes of farming or harvesting the earth. One, the western capitalist version that seeks to exploit agriculture for economic ends; the other, the Indigenous approach that, as the artist Patricia Domínguez has recently explained, functions as a 'vegetal matrix', a way of 'conceiving plants as companions, organic technologies [...] that have provided us with multi-species connection' (Domínguez, 2022: 13).

What is ultimately at stake in this conflict is a dispute over two ways of inhabiting the earth. The first – the peyote or Indigenous one – is an attempt to 'alter' the human, in the full sense of that word, to open it out, physically and metaphysically, to terrestrial and cosmic forces, to provoke a poesis of connection. The second, by contrast, is intended to keep humans at a distance from the world, to veil the realities of death and suffering, to separate species into those deemed worthy of life and those whose fate is to serve, to be sacrificed. Artaud does not, then, wager on Tarahumara technology because it is more organic or 'natural' – peyote is a hallucinogen, after all – but because it is more creative and life-affirming than the technologies involved in capitalist agriculture. Peyote eating is a mechanism that discloses a future; it resists death and despoilation. Not only does it allow the world to be reborn differently, but, for Artaud, it calls out for a new planetary populace willing 'to renounce the carapace of the present world [...] torn apart by ten thousand wars, and evil, and disease, and poverty, and the scarcity of provisions, objects, and substances of the first necessity' (Artaud, 1976: 518).

Although he participated in the Tutuguri or peyote ritual, Artaud did not seek to become one with the Tarahumaras, and neither was he concerned to draw on their rituals dramatically in the manner, say, of Peter Brook, Jerzy Grotowski or Richard Schechner.[32] Anticipating the 'returns' to Indigenous practices that many contemporary environmental thinkers have made in the twenty-first century, he is drawn to them because he realises – as early as the 1930s – that they offer the most pragmatic of planetary choices, a line of flight from the insanity and suffering of western notions of progress that look to exterminate everything that would prevent the ontotheological quest for oneness and total profit.[33] That the 'preference' Artaud has for the Tarahumara peyote priests is much more than a reactionary primitivism, a private attempt to recover some lost authenticity, is well explained by the ideas of Philippe Pignarre and Isabelle Stengers in *Capitalist Sorcery: Breaking the Spell* (2011), a poetico-polemical text informed by Tobie Nathan's contemporary ethnopsychiatric work with diverse African communities in Paris.

In *Capitalist Sorcery*, Pignarre and Stengers contend that western capitalism imposes a hegemonic system of thought and value, a rigid onto-epistemology that condemns the planet to conform to its spurious, utilitarian logic of good sense and disenchanted materialism. To bolster this system, to universalise its abstraction, any challenge to it, such as Artaud's attraction to magic, alchemy and peyote, is immediately shut down, regarded, at best, as idealistic, and, at worst, an instance of dubious primitivism. But, as Pignarre and Stengers demonstrate, western rationalism's recourse to 'truth' rather than 'superstitious belief' does not absolve the world of magic; it serves only to create a more perfected and pernicious system of sorcery that hides itself under the name of 'proper' science, a type of ruthless witchcraft that allows the 'spirit of capitalism' to impose itself as the only possible way of existing in the world (Pignarre and Stengers, 2011: 64). The logic of the sorcerer is terrible and perverse: they accuse their enemies of the very crimes they commit:

> To dare to place capitalism in the lineage of systems of sorcery is not to take an ethnological risk but a pragmatic one. Because if capitalism enters into such

a lineage, it is in a very particular fashion, that of a system of sorcery without sorcerers [...], a system operating in a world which judges that sorcery is only a simple 'belief', a superstition that therefore doesn't necessitate any adequate means of protection. (Pignarre and Stengers, 2011: 40)

Neither critique nor activism is enough by itself to escape capitalism's 'sorcery without sorcerers', Pignarre and Stengers contend. The more efficacious method – and this is precisely what Frédéric Neyrat argues in his brilliant revisionist account of Artaud's colonial politics in *Instructions pour une prise d'âmes: Artaud et l'envoûtement occidental* (2009) – is to confront capitalist sorcery with a different kind of magic, something that looks to exorcise demons, combating depression with life, wagering on a new kind of enthusiasm. As Pignarre and Stengers proclaim:

> We refuse to share this pedagogical faith in Truth and in a future that is as promising as the present is sinister. We see in this a way of dying with dignity, without renouncing the ideal that transcended the life of those who adhered to it: a *modus moriendi* and not a way of living and creating, a *modus vivendi*. For our part we want to think against the fairy tale of progress, with or without the Marxists. And to do that, it is necessary to learn not only to yearn but also to learn fright in the face of the power of the operation of capture that has made us so proud of our 'sober senses'. (Pignarre and Stengers, 2011: 61)

To read Artaud alongside Stengers and Pignarre is to disclose an anticolonial energy in his thinking that other critics have overlooked.[34] Artaud is not interested in using the Tarahumaras to revive western culture from without, to save it from decadence and collapse. What concerns him is the possibility of inhabiting a different world, predicated, as I argue on pp. 164–72, on animism and alchemy as technologies for contesting the catastrophes of capitalist sorcery.

Despite undergoing more than one hundred electroshock and insulin treatments, and even when suffering from the agonies of stomach cancer and heroin withdrawal in the late 1940s, Artaud never lost faith in the capacity of magic to recreate the world for the purposes of a healthier, more affirmative mode of existence. Antedating the work of contemporary postcolonial thinkers such as Achille Mbembe, Artaud, in his late work, continued to excoriate western humanism as a death cult, an ontotheology whose obsession with property and profit margins 'no longer support[ed] life' (Artaud, 1958: 8). In the 1947 essay 'Van Gogh: The Man Suicided by Society', this necropolitics is associated with capitalism's destructive attempt to discipline individual life by subjecting it to continual administrative appraisal and control – a type of biopolitical pressure that results in the deconstruction of the individual organism through a kind of spectral 'suiciding', the insinuating work of a capitalist sorcerer:

> One does not commit suicide by oneself.
> No one has ever been born by oneself.
> No one dies by oneself either.
> But in the case of suicide, there must be an army of evil beings to cause the body to make the gesture against nature, that of taking its own life. (Artaud, 1976: 511)

In the opening section of *To Have Done with the Judgement of God*, written in the immediate aftermath of the implementation of the US Marshall Plan in western Europe and Japan in 1947–48, this same nihilistic denial of life manifests itself historically and collectively, in the insane proliferation of chemical pesticides, industrial agriculture and artificial insemination and nuclear weapons during the Cold War. Two decades before Rachel Carson compared the widespread use of DTT to the dangers of nuclear fallout in the environmental bestseller *Silent Spring* (1962), Artaud is aware that advanced capitalism is a *modus moriendi*, a technology of planetary despoliation, an economics of devastation. I cite at length the opening section of *To Have Done with the Judgement of God* in order to give some indication of the sophistication of Artaud's ecopolitics, his awareness that piecemeal responses of the sort demanded by environmental reformists and sustainable developers is not enough. Only a complete abandonment of capitalist modernity will do:

> Because there must be production
> Nature must be replaced wherever it
> Can be replaced by every possible
> Manner of activity,
> A major field must be found
> For human inertia,
> The worker must be kept busy
> At something,
> New fields of activity must be created,
> Where all the false manufactured
> Products,
> All the vile synthetic ersatzes will
> Finally reign,
> Where glorious true nature will just have
> To withdraw,
> And give up its place once and for all
> And shamefully to all the triumphant
> Replacement products,
> Where the sperm from all artificial
> Insemination factories
> Will work miracles
> To produce armies and battleships.

No more fruit, no more trees, no
More vegetables, no more ordinary
Or pharmaceutical plants and consequently
No more nourishment,
But synthetic products to repletion
In vapors,
In special humours of the atmosphere,
On particular axes of atmospheres
Drawn by force and by synthesis from
The resistance of a nature that has
Never known anything about war
Except fear.

(Artaud, 1992: 311)

To situate Artaudian cruelty within these environmental, technological and militarist histories is to reanimate it with intensified ecological life, to contest one-sided claims about its nihilism and primitivism. Implicit within the aesthetic 'bombs' that he hurls at the world, there is a clear clinical and political impulse at work in Artaud's ecology that contemporary spectators would do well to attend to. Artaud's aim was never just to consume the cosmogonies or performance traditions of other cultures as a way of personal salvation. Rather his primary purpose was to target the ecopolitical damage inflicted by the historical monologue of 'witchcraft Europe', to contest, like his contemporaries Theodor Adorno, Walter Benjamin and Max Horkheimer, its irrational rationality, the suffering that its capitalist humanism has provoked for so many peoples and species on the planet.

I make this point with such passion because it provides a context that Jacques Derrida is notably silent about in his 'powerful' deconstruction of Artaud's theatrical project. Since Derrida's readings have done so much to determine how Artaudian cruelty has been perceived in Theatre and Performance Studies from the mid-1970s onwards, one is required to return to his readings in some depth. There is an additional imperative in this return. For what is so profoundly at stake in Derrida's hesitation to endorse, fully, Artaud's project is not just a disagreement about theatre's role in the production of metaphysical presence – performance without repetition – but, more radically, its capacity to function as an ecopolitical apparatus. As I explain below, Derrida's suspicions about cruel theatre mirror his aggressive refusal to countenance the possibility of an Artaudian ecology. In both instances, Derrida contends that any straightforwardly positive appropriation of Artaud's ideas, be that dramatic and/or environmental, runs the risk of condoning a reactionary politics in which difference is eradicated and where, as I intimated in my own reading of Corvin's abhumanism, a destructive humanism returns in inverted form.

Liberating Artaud's ecology

In his essays on cruel theatre in the mid-1960s, Derrida starts by cautioning against simplistic readings of Artaudian cruelty that would merely emphasise its gratuitousness:

> Artaud does not call for destruction, for a new manifestation of negativity. Despite everything that it must ravage in its wake, 'the theatre of cruelty' is not the symbol of an absent void. It affirms, it produces affirmation itself in its full and necessary rigour. (Derrida, 1978: 292–4)

In Derrida's parsing, Artaud's attack on representation (repetition) goes beyond aesthetics. Its real ambition, Derrida recognises, is to undermine the metaphysical framework of western thinking itself, with its logocentric attachment to 'an author-creator who, absent and from afar, is armed with a text and keeps watch over, assembles, regulates the time or the meaning of the representation' (Derrida, 1978: 296).

As the logocentric art form par excellence, Derrida understands, like Artaud, that western theatre is paranoid, monologic, jealous about maintaining its dominion over the 'globe'. Unless its underlying 'logic' is actively contested and counter-actualised, no change is possible. Replacing one message with another is not enough. The coherence of the *logos* itself is what needs to be targeted:

> This general structure in which each agency is linked to all the others by representation, in which the irrepresentability of the living present is dissimulated or dissolved, suppressed or deported within the infinite chain of representations – this structure has never been modified. All revolutions have maintained it intact, and most often have tended to protect or restore it. (Derrida, 1978: 297)

Perhaps predictably for a deconstructionist, it is at the very point when Derrida appears to touch on the 'profound essence' of Artaud's project that he insists on maintaining a distance, in positing a limit to his affirmation (Derrida, 1978: 310). For though Derrida certainly shares Artaud's critique of the *logos*, he also maintains that the latter's desire to do away with repetition is an error, a 'fatal' flaw: 'Theater as repetition of that which does not repeat itself […], such is the fatal limit of a cruelty which begins with its own representation' (Derrida, 1978: 316). Behind Artaud's seemingly aggressive rejection of God, there remains, Derrida contends, a theological structure in operation, a burning desire, on Artaud's part, to speak in his proper voice, to be whole, substantial. Ironically, then, in its attempt to inflict cruelty on the *logos*, 'Artaud's "metaphysics" fulfils', Derrida proposes, 'the most profound and permanent ambition of Western metaphysics': namely, 'unity prior to dissociation', the impossibility of 'life without difference' (Derrida, 1978: 244).

I mention all of this not to go over old ground but to provide some context for grasping Derrida's little-discussed caution about reading Artaud ecologically. In the long essay 'Artaud the Moma', delivered initially as a lecture at the Museum of Modern Art (MoMA) in New York in 1996,[35] Derrida confidently proclaims that in order to remain true to Artaud's 'lightning strike' and 'thunderbolt', he will:[36]

> [R]esist everything in this work that, in the name of the proper body or the body without organs, in the name of a reappropriation of self, is consonant with an ecologio-naturalist protest, with the contestation of biotechnology, reproductions, clones, prostheses, parasites, *succubim*, supports, specters, and artificial inseminations – in short everything that is im-proper and that Artaud-Mômo, as you heard, identifies very rapidly with America, in 1947, in *Pour en finir avec le jugement de dieu* (*To Have Done with the Judgement of God*). (Derrida, 2017: 6)

Uncharacteristically, what Derrida means by this provocative dismissal of Artaud's ecology (and oblique attack on Deleuze and Guattari) in *To Have Done with the Judgement of God* is never properly clarified or unpacked. Once stated, Derrida never comes back to it, at least not explicitly. What is evident beyond any doubt, however, is that Derrida is unwilling to accredit Artaud's prescient critique of the US pharmaco-chemical, military-industrial complex in that text with any real seriousness, seeing it only as an index of some Rousseauian desire to return to a proper idea of 'Nature' without prosthesis or artifice – a move that mimics, almost entirely, his earlier reading of Artaud's theatre. When it comes to ecology, Artaud is, once again, deconstructed by Derrida for attempting to recover an impossible, reactionary origin that runs the risk of an authoritarian politics, in which everything im-proper is rejected for a fantasy of origins, similitude and essence – the same thing that motivates Jannarone's critique of his work.

But what if Derrida has made a mistake here, a kind of category error? Could it be that, in his impassioned quest to save Artaud from a logocentric episteme, he gets the historical and ontological dimensions of his ecology wrong? And if this possibility is accepted, how then to emancipate Artaud's ecology from Derrida's misunderstanding so that the environmental politics of cruel theatre are made visible? While he makes no reference to Artaud, Michael Marder's critical take on what he calls Derrida's 'allergy to ecology' offers a way forward:

> I would go so far as to say that Derrida is allergic to 'ecology'. Usually alive to the excesses of metaphysics, he becomes hypersensitive (and what is allergy if

not hypersensitivity?) and overreacts to this word together with everything it articulates and signifies. (Marder, 2018: 143)

According to Marder, Derrida's allergy to ecology in his writing, along with his reluctance to embrace environmentalism as a positive and progressive force, is caused by his failure to read the *oikos* as anything other than a signifier of home, a prefix for a metaphysical economics in which 'Nature' is self-coincident with itself, something proper and proprietorial:[37]

> What Derrida is (doubly) allergic to [...] are the constituents of ecology: the familial dwelling (*oikos*) and, above all, the trenchantly untranslatable logos that gathers into itself voice, speech, and the very act of gathering, all of them mercilessly ground in deconstruction's interpretative mill ever since the first writings on the phenomenology of Edmund Husserl. Given the way Derrida treats the two ingredients of this composite term, it is safe to assume that nothing in this attitude would change once they were put together, their force doubled. (Marder, 2018: 162–3)

Intriguingly, Marder's critique of Derrida's aversion to ecology is rehearsed, but in a more theatrically focused and Artaud-specific context by the philosopher Howard Caygill. Where Marder speaks of allergens, Caygill uses the related metaphor of auto-immunity to describe Derrida's tense relationship with Artaud's theatre. For Caygill, Derrida's 'Artaud-immunity' is motivated by a fear of being 'infected', the realisation that his own way of doing philosophy could be compromised by the cruel affects and sensations championed by Artaud:

> For Derrida, Artaud posed the problem of how to read an *oeuvre* whose defenses had been pre-emptively disabled in an auto-immune reaction to anticipated rejection that seemed to verge on the suicidal. [...] The limits of Derrida's understandings of auto-immunity are set by his Artaud-immunity that underestimated Artaud's *counter-resistance* as a chaotic Platonism in which the lowering of defenses was anything but suicidal. (Caygill, 2015: 114; original italics)

Placed together, Marder and Caygill point to a possible cause for Derrida's misreadings of Artaud's theatre and ecology: the fact that he was unable to think beyond his own philosophical system and allow for the possibility that Artaudian cruelty waged on a different understanding of 'Nature' that had little in common with his own concepts. Instead of taking Artaud's ecology seriously, Derrida seems to have inoculated himself by imposing his own paradigm of thought upon it. As a consequence, he could only ever approach Artaud as a dangerous allergen, a virus that had to be guarded against, kept at a distance, through a fear of being contaminated, and thus brought to the brink of philosophical suicide. How then to reverse this misunderstanding? What can be done to liberate Artaud's ecology from the quarantine in which Derrida has placed it, and which other critics have tended to repeat?

A useful way of circumnavigating Derrida's 'Artaud-immunity' is to approach Artaud's ecology *as* ecology and not as an inverted allegory for a metaphysical desire to be whole, some oedipal fantasy to be his own 'mother, father, and son', giving birth only to himself, as Derrida proposes (Derrida, 1978: 315). In his nuanced but remarkably consistent critiques of Artaud's commitments, Derrida starts, always, with the assumption that what Artaud wants is to institute a proper way of communing with the earth, to return to a primal ground, existing as a kind of 'deep ecologist' *avant la lettre*, someone who, as I mentioned in Chapter 2, seeks recovery and repair.[38] But Artaud was never a 'deep ecologist'. At all times, he is what I have been calling a 'theatre ecologist'. Artaud's ecology has little interest in restoring presence or living in harmony with 'Nature' in any kind of holistic way. Well before Derrida came on the scene, Artaud is aware that self-presence is impossible and that 'Nature' is cruel, always splitting and diverging, a dynamic, individuating 'chaosmos' that humans are part of and simultaneously yet separated from.

In marked contrast to what Derrida claims, difference, for Artaud, is not naively imagined as an otherness or alterity to be overcome through a dialectical negation, a moment of sublation; it remains stubbornly persistent as the force that drives becomings, an expressiveness within matter that escapes any attempt to capture or harness it for human ends alone. As I explain in greater detail in the section on alchemy and animism in this chapter, Artaud contends that difference is immanent to earthly life, a diffractive, individuating potential that every material organism, both organic and inorganic, contains within itself.

From this perspective, Artaud's materialist understanding of difference conflicts, irreparably, with Derrida's celebrated grammatological definition, based on the deferral of presence that writing allows for. In *Margins of Philosophy*, Derrida defines difference (*différance*) as a 'finality without end', an infinite or messianic movement that can never die, precisely because it has no terrestrial origin to account for its appearance (Derrida, 1982: 5). As this comment demonstrates – and this is even more pronounced in his later reflections on messianic temporality – difference appears to arrive from nowhere, a space beyond the earth, the linguistic gift of a radically transcendent, messianic God. In his predilection for a messianic signifier whose 'to come' (*l'avenir*) is endlessly deferred, Derrida forgets that human existence is rooted on a planet and in life forms that change, evolve *and*, crucially, 'come to an end', as Claire Colebrook has argued in a salient ecological critique of Derridean grammatology (Colebrook, 2018: 263). To live on the earth is to accept, as Lyotard always does, that the planet will eventually explode, leaving no remainder in its wake, no ghost, trace or sign.

Because he is so concerned to keep difference ungrounded and messianic, so free of an earthly origin, Derrida fails to give credence to the ecological affects of Artaudian cruelty. As Frédéric Neyrat notes, for Artaud, as opposed to Derrida, there is something 'innate' ('un inné') to earthly existence that is 'neither nature, nor culture' ('ni nature, ni culture'), a pre-individual energy that all beings contain within themselves but which it is impossible to coincide with or make present (Neyrat, 2009: 68; my translation). The most – the best – one can do with this innate force is to follow its process, to assent to the incomplete 'birthings' that it provokes.[39]

Although theatricality, as I suggested in Chapter 4, prohibits any immediate or absolute encounter with this terrestrial power, it does not simply pass it along a chain of phonemes and semantemes, as the messianic play of Derridean linguistics maintains. For the innate is not to be reduced to meaning or truth, a transcendent communion between subject and object; it is an anarchic metastable energy flow that is felt through a kind of 'inflammatory participation' working through 'the senses and not the logos' (Caygill, 2015: 130). This flow is always on its way elsewhere, a furtive and elusive 'double' that no word can reflect nor body figure. Ultimately, the innate belongs to 'another ontology' ('une autre ontologie') (Neyrat, 2009: 68; my translation). In this terrestrial ontology, there is nothing to recover and neither does the call to affirm existence, to live finally, come from some messianic spectre, a voice without grain or flesh that knows nothing of finitude and endings, as Derrida suggests. For the innate, as I explained in my parsing of Simondon in Chapter 2, is an im/material surplus that emerges from and belongs to the earth itself, something sublunary, an immanent quality of matter that will disappear when the earth undergoes its future heat death in five billion years from now. The innate, like Simondon's pre-individual, can only become; it can never *be*. Within the (deep) timespan of the earth's life cycle, its destiny is to change, mutate and multiply, to variegate. It is experienced as a quality not an object, a haecceity, a fullness not a lack.[40]

In opposition to what Derrida claims in his writings on Artaud, neither theatre nor 'Nature' is a site for reconciliation, absolution and self-presence.[41] Instead, to return to Simondon's idiom, they are vehicles of/for terrestrial individuations, material assemblages that impress themselves on bodies and open up virtual possibilities through the expression of intense affect. The innate is never brought to completion or actualised; it operates through partial and theatrical 'cuts' that tie and untie at the same time, the very 'genesis of creation' as Artaud has it (see pp. 147–8). In Artaud's thinking, the subject experiences a million births during its stay on earth.

The tragedy of capitalist sorcery is to have convinced humans that there is only one birth, a singular identity that is granted to us, an origin that we are compelled to restore, to remain faithful to. Reconsidered from this perspective, there was no need for Derrida to have protected Artaud from what he calls 'ecologio-naturalist protest'. The theatrical 'limit' that Derrida insists on salvaging in his interpretations of Artaud is always already at work within ecology itself, as the disjunctive-conjunctive relation that produces terrestrial changes and earthly becomings. This limit is not the outcome of a 'flawed' metaphysical project, a sad victory in spite of itself, as Derrida's deconstructive readings of Artaud insinuate.[42] Rather, it is joyously and intentionally desired by Artaud as a necessary factor for a 'different way of thinking difference' (Cull, 2009: 244). One that insists on a theatrical 'out of stepness', a terrestrial stage in which bodies and things are transformed by cruel energies that bring them together but also tear them apart and where western philosophy's identity project is foreclosed in advance. In this sensate and materialist concept of difference, theatricality is a virtuality, a prodigal line of light, an ex-orbitance of the inside – an intensity, radically non-binary from the outset. It has no need of deconstruction.

In a provocative reversal of Derrida's critique of Artaudian corporeality, only a theatre that insists on making bodies appear in their 'presencing' can ever hope to escape the straight jacket of presence (be that theological or ecological). In contradistinction to Derrida, Artaud does not see theatre as a philosophical stage, a place to deconstruct the signifier in the name of some infinite act of hermeneutic undecidability. It is cosmic-terrestrial, an affective *plateau* where variation holds sway and a new kind of alchemy happens, a site where the subject experiences what Artaud, at the very end of *To Have Done with the Judgement of God*, was the first to call 'a body without organs' (Artaud, 1992: 329).[43]

As this reclamation of Artaud proposes, an ecologically cruel theatre is not one that does away with representation in the hope of rediscovering a lost essence, a pristine or 'deeper' way of being with 'Nature'; it is one that allows the bodies of spectators to be touched by the innate expressivity of matter, placing them in proximity to a life force that is *beyond*, *between* and yet also already *within* them. Artaud's ecologically primed body is an organ-less body, never wholly flesh, energy or form, but always actual *and* virtual, matter that thinks, an aggregate of stuff that lives, dies and transforms. As a means of exploring, in greater detail, the full ecological significance of Artaud's cruel theatre, this *place* which, as I have stated, is a *plateau*, two key concepts remain to be thought through: alchemy and animism. For, as ever with Artaud, accepted definitions are subjected to a singular rerendering, an ecological recalibration that the ideas and vocabularies of Barad, Lyotard and Guattari help to illuminate. If indeed, 'illumination' is the correct word for something that remains irrefutably opaque, located, as it is, within what Artaud defines in the 'Theatre of the Seraphim' as 'the walls of some underground tunnel' – the same shadowy 'cave space' that troubled Plato (Artaud, 1976: 274–5).

Alchemy and animism

In the 1932 text 'The Alchemical Theatre', Artaud starts by advancing what seems to be a somewhat conventional, metaphorical relationship between theatre and alchemy predicated on a shared interest in the metamorphosis of base material into something more beautiful and complex.[44] 'In books dealing with alchemical subjects', he proposes, 'there is a curious predilection for a theatrical vocabulary' (Artaud, 1958: 49). However, as the essay proceeds, Artaud excavates a 'deeper resemblance' between these practices that, for him, is found on an ontological or molecular level. This 'inner articulation', this point of queered indiscernibility, is found in how both alchemy and theatre, in their refusal to separate formlessness from form and mind from matter, work to express the innate play of what Artaud terms the 'essential drama' (Artaud, 1958: 51).

In keeping with the biochemical cosmogonies of early modern alchemists such as Paracelsus, Heinrich Kunrath and Elias Ashmole, material reality, for Artaud, is subtended by an invisible principle – an innate will to become – that if skilfully organised can produce new forms of life. A point that is underscored by Ann Demaitre in her insightful article on Artaud and alchemy from 1972:

> In search of the ultimate physical and spiritual reality, the alchemist relied on the Platonic assumption that behind visible, tangible things lies a spiritual essence, 'the world-creating spirit concealed or imprisoned in matter'. In the hermetic language of the alchemists, this spirit is referred to by a variety of names such as Essence, Pneuma, Mercurius, etc. But whatever the name and its hidden significance, in the alchemists' view the spirit entombed in matter was the creative element, 'the power of the whole cosmos from the fixed stars to the center of the earth'. (Demaitre, 1972: 240)

Artaud's theatre both confirms and dissents from Demaitre's explanation. It confirms it by adhering to the ontogenetic capacity that alchemy attributes to matter, and dissents from it in the comparison that Demaitre draws with Plato. For while Artaud certainly believes that there is a 'spiritual force' residing with or 'behind, visible tangible things', as Demaitre suggests, he does not equate this force with the transcendent, immutable realm of Platonic ideas. As I have already stated, innateness is anti-Platonic: it belongs to the shadow-play of theatre, cave and tunnel, sites in which inorganic and organic bodies get confused, shape shift, becoming other than what they are. Alchemy, according to Artaud, is learning to assent to this theatro-terrestrial 'mutancy', to the fact that life is a process of diffractions and cuts that all organisms participate in, a properly virtual art.

Mind is not simply *in* matter, for Artaud, an external principle or substance that could be potentially separated from the stuff of the world. Rather mind *is* matter, an inalienable property of existence, an agentic force. In a key passage mid-way through the 'The Alchemical Theatre', he explains how:

> [T]his essential drama, we come to realise, exists, and in the image of something subtler than Creation itself, something which must be represented as one Will alone – and *without conflict*.
>
> We must believe that the essential drama, the one at the heart of all the Great mysteries, is associated with the second phase of Creation, that of difficulty and of the Double, that of matter and the materialization of the idea.
> (Artaud, 1958: 51; original italics)

In Artaud's alchemic ontology, the universe is composed of a single sheet of matter-energy, belonging to the chaos state of the 'first time of creation' – the time that, on the surface at least, *appears* to lack 'difficulty' or complexity. On this molecular plane, everything is nebulous and amorphous, a biochemical flux, a oneness. By contrast in the 'second phase of creation', that same amorphous flux is now endowed with form, an operation that seems to replay Aristotle's decidedly anthropocentric notion of hylomorphism.[45] In what Artaud refers to as 'the time of the Double', structures start to emerge and individuation takes place – the 'materialisation of the idea'. Without the Double, Artaud says, 'there can be no theatre no drama' (Artaud, 1958: 51). For, as Artaud reminds the reader, life – 'true theatre' – is 'organised chaos', the consequences of 'primitive unifications' that the Double gives birth to. Importantly, though, the molecular anarchy that the Double appears to organise is never fixed or solidified, a primal stage that has been left behind. Rather, the world formed by the Double remains 'a cosmos in turmoil', a metastable world that is liable to transform and become chaos again (Artaud, 1958: 51).

Artaud's insistence on metastablility is revealing. Despite his normative talk of 'firsts' and 'seconds' and 'originals' and 'doubles', in Artaud's thought chaos and cosmos are neither temporally sequential nor spatially opposed. Conversely, these two times of creation are folded into and through each like a Borromean knot, caught up in a relationship that Artaud is at pains to characterise as 'more subtle' than 'Creation itself' (Artaud, 1958: 51). Against the logic of linear causality, the Double, in Artaud's cosmogeny, is always already at play from the beginning, haunting the first time of creation with the ghost of futural possibility, an agentic force or im/materiality that undoes all oppositions in the name of multiplicity and potentiality. The Double, then, does not reflect a spiritual reality nor infuse the chaos with a pre-existing set of ideas from without, as Demaitre posits and Aristotelian hylomorphism assumes. Instead, 'the Double', to cite the Artaudian scholar Eric Sellin, is 'the Double of itself', an origin without an origin, a subtlety so complex and diffractive that no thought or word can represent or hold it (Sellin, 1968: 94). Strange as it may seem, the Double does not come after the first time of creation; it arises at the same time. The Double is present in chaos, a kind of form in formlessness. Put in Simondon's terms, the Double is a manifestation of the pre-individual, an affect not a thing, a molecular energy flow of 'potentials and virtualities' that are inherent to matter itself, even as it appears to come after it (Simondon, 2017: 253).

The conjunctive–disjunctive relationship that defines both alchemy and theatre accounts for why Artaud, later in the essay, is careful to talk not of dialectics *per se*, but of an 'almost dialectical sequence' (Artaud, 1958: 49). That is, a dialectics without synthesis, an impossible relationality in which forms and forces, as well as signs and things, simultaneously connect and separate from each other, existing in what Artaud calls 'unimaginably strange conjunctions' that 'strip man of his head' (Artaud, 1958: 49.) *Contra* both Plato and Derrida, Artaud already knows, like Mallarmé, that 'there is no present', and that difference precedes being (Mallarmé, 2018: xii). Such is the cruelty of the 'essential drama', the conflict that can never be resolved or sutured, the 'heart' of what Artaud calls 'the Great Mystery' (Artaud, 1958: 51).

According to Artaud, if it is to impact on life, theatre has to construct a dramaturgy capable of staging this alchemical 'severing together […] in a completely physical way' (Artaud, 1958: 49). At their isomorphic core, theatre and alchemy are the same because they organise chaos, make anarchy productive, and so release the 'more than plenitude' of the Double. The skill in both operations is to reach the subtractive point that will allow transformation to take place, without, on the one hand, domesticating the chaos of the pre-individual flux, and/or, on the other, allowing chaos to defeat form and so dissolve into amorphousness. Both must be held in a state of 'chaosmotic' tension, assembled as a diffraction pattern, and shown to be fundamentally theatrical, that is to say, indeterminate, oscillating, 'more than one'.

Ultimately, it is this excessive relationship, this staging of doubleness, that defines Artaud's theatre as posthumanist rather than abhumanist, to return to Corvin's language. The 'abstract purity' that Artaud is after is not an attempt, like Craig's Über-Marionette, to locate a new wholeness beyond the divisions of the human. It designates an indiscernible *stage* (or station) in an infinite and erratic adventure, one in which the ambiguous and ambivalent reality of the Double produces endless becomings and mutations. 'This reality', Artaud insists, 'is not human but inhuman, and man with his customs and his character count for very little in it' (Artaud, 1958: 48). Everything on this organless, abstract *plateau* is in process, actual *and* virtual at the same time, partial *and* full. There is no going beyond this plane because there is nowhere else to land, no stable ground to attain, nothing behind the Double. All one can do is to to be affected by it, to learn how to distil it:

> The theatrical operation of making gold, by the immensity of the conflicts it provokes, by the prodigious number of forces it throws one against the other, by this appeal to a sort of essential re-distillation brimming with consequences and surcharged with spirituality, ultimately evokes in the spirit an absolute and abstract purity, beyond which there can be nothing, and which can be conceived as a unique sound, defining note, caught on the wing, the organic part of an indescribable vibration. (Artaud, 1958: 51–2)

In theatre's alembic, the voice of the *logos*, the theological one that Derrida thought he heard in his readings of Artaud, is silenced. What emerges in its place, is a 'unique sound', the sonic expression of a disruptive frequency, an 'abstract purity' that emanates from and courses through a cosmic earth. The noise of this indescribable vibration, this 'hum', stretches everywhere and nowhere, a deep, blank sonority that fills space without ever being reducible to that space, primarily because it is always moving elsewhere. In the thought of Deleuze and Guattari, cosmic alchemy exists on a plane 'where no gods go' and where discrete objects, forms and signs are replaced by 'forces, produced intensities, vibrations, breaths' – a multiplicity of half-tones, frequencies and rhythms (Deleuze and Guattari, 1987: 158).

For western theatre to connect with this 'plane of consistency', to liberate the heterogeneous, embryological voice of the 'pre-natal' (Artaud, 1976: 499), Artaud demands that the written text – the script – should lose its privileged position:

> The library at Alexandria can be burnt down. There are forces above and beyond papyrus; we may temporarily be deprived of our ability to discover these forces, but their energy will not be suppressed [...]. It is right from time-to-time cataclysms occur which compel us to return to nature, i.e. to rediscover life. (Artaud, 1958: 10)

Thinking, for Artaud, does not take place in the mind alone, as the dominant idealist or correlationist trend in western idealist philosophy suggests. On the contrary, it occurs in conjunction with elemental forces outside itself, in waves of matter-energy breaking on top of and seeping into bodies:

> Perhaps even man's head would not be left to him if he were to confide himself to this reality – and even so it would have to be a stripped, malleable, and organic head, in which just enough formal matter would remain so that the principles might exert their effects in a completely physical way. (Artaud, 1958: 48–9)

Instead of assenting to this non-human, molecular reality, western modernity, Artaud says, has a gaze turned inward, a narcissistic compulsion to turn away from the elementalism of the earth, to 'stop the idea of the world' by wagering 'on the infinitesimal inside' (Artaud, 1992: 319). The consequence of this inward gaze has been disastrous environmentally and politically. Prefiguring the decolonial ecologies of Glissant, Yusoff and Ferdinand, it has resulted in the triumph of 'whiteness', the colour of death, the loss of all *savoir*: 'As iron can be heated until it turns white, so it can be said that everything excessive is white; for Asiatics white has become the mark of extreme decomposition' (Artaud, 1958: 9).

Alchemical theatre, for Artaud, is intended to overcome this unskilled, colonialist metallurgy by preparing aesthetic forms that will allow spectators 'to touch life' (Artaud, 1958: 13). The goal of alchemy in performance is not to create a holy theatre that would take us beyond the earth, as so many theatre scholars have thought. Rather the intention is to approach life itself as something sacred, a transformation in perception that posits theatre as an animated and animistic site, a place for magical mutations. Once again, and to return to my argument on pp. 149–57 in this chapter, it would be a mistake to regard Artaud's 'animism' as some naive appropriation of the 'primitive', an act of cross-cultural piracy. He seeks rather, as indeed do contemporary ecological thinkers such as Stengers and Latour,[46] to deal with a mode of agentic agency that western science has consciously turned its back on, but which it has never been able to properly surmount, despite its efforts to do so: the very real presence of an autonomous, lively earth, the same one that caused the Lisbon earthquake of 1755, ruining the Enlightenment before it had even happened, foreclosing its claims.[47] As with Stengers, 'reclaiming animism' is a pragmatic operation, for Artaud. It means thinking in 'terms of assemblages that generate metamorphic transformation in our capacity to affect and be affected – and also to feel, think, imagine' (Stengers, 2012: 9).[48] Or, to use Artaud's terminology, it is to attune oneself to 'the expression of a primary Physics from which the Spirit has never separated' (1958: 60), a vitalist 'metaphysics' (Artaud, 1958: 38) that would 'allow us to believe in what makes us live and that something *makes* us live' (Artaud, 1958: 7).

Ecological light is shed on Artaud's 'animated sorcery', by reading it in tandem with Barad's ideas (Artaud, 1958: 73). In a paragraph that resonates with Artaud's thought, Barad contends that life is a passionate, vitalist affair, an instance of quantum dis/continuity in which every repetition is a creation, an intervention into the real, a way of being touched:[49]

> The point is not merely that something is here-now and there-then without ever having been anywhere in between, it's that here-now, there-then have become unmoored [...]. A rupture of the discontinuous? A disrupted disruption? A stutter? A repetition not of what comes before, or after, but a disruption of before/after. A cut that itself is cut across. A cut raised to the higher power of forever repeating. (Barad, 2010: 247–8)

With Barad as quantum companion, it becomes possible to see how Artaud's alchemical theatre really is a 'primary physics', an apparatus endowing the world with a multitude of virtual possibilities in the here and now. In it, bodies are molecular, intra-active and de-phased, both particle *and* wave, doubled doubles:

> There are no separately determinate individual entities that interact with one another; rather, the co-constitution of determinately bounded and propertied entities results from specific intra-actions [...] which break [...] open the binary of stale choices between determinism and free will, past and future. (Barad, 2010: 253–4)

As vehicle for this monadic, intra-active flux, theatre does not need to stand for something, to point to a reality beyond itself. Its doubleness is not reflective. Rather its signs are real without being true, animated events that provide access to 'those incandescent futures' which Artaud was so attracted to. Coterminous with what Barad calls 'agential cuts' that effect 'a holding together of the disparate itself' (Barad, 2010: 265), Artaud's alchemical stage is leaky, ambidextrous, diffractive. It is both separated from *and* joined to a material world that it intervenes into, haunting it with fleshy ghosts that straddle both realms, impossibly, at the same time. Or, to have recourse to one of Artaud's most celebrated maxims, 'If theatre doubles life, then life doubles true theatre' (Artaud, in Derrida, 1978: 433, n.19).

Artaud's attempt to express the virtuality of this doubled life – to stage it – explains his predilection for shadows, those impossible signifiers that partake of the opaque force that theatricality brings to the surface of bodies:

> Like all magic cultures expressed by appropriate hieroglyphs, the true theatre has its shadows too, and, of all languages and all arts, the theatre is the only

one left whose shadows have shattered their limitations. From the beginning
one might say its shadow did not tolerate limitations. (Artaud, 1958: 12)

Like Barad's diffractive light waves (see pp. 106–10), Artaud's theatrical shadows teem with possibility and life. No beam is strong enough to illuminate their animated play, to exhaust their enigma, to stop their becoming. To put a body on stage is to 'show' opacity, to evoke an obscure realm of existence inhabited by the virtual, a fugacious and productive force, 'a necessity that cannot *not* create' (Artaud, 1958: 102; original italics):

> But the true theatre, because it moves and makes use of living instruments, continues to stir up shadows, where life has never ceased to grope its way. The actor does not make the same gestures twice, but he makes gestures, he moves; and although he brutalizes forms, nevertheless behind them and through their destruction, he rejoins that which outlives forms and produces their continuation. (Artaud, 1958: 12)

As Artaud's celebration of the generative power of gesture intimates, theatre's capacity to 'stir up shadows' is dependent on gratuity, on the willingness of the performance to accomplish nothing, to destroy what it has created, to undo what it has brought into being. Once this gestural economy is accepted, a different kind of possibility announces itself. In a celebrated passage, Artaud notes how:

> Once launched upon the fury of his task, an actor requires infinitely more power to keep from committing a crime than a murderer needs courage to complete his act, and it is here, in its very gratuitousness, that the action and effect of a feeling in the theatre appears infinitely more valid than that of a feeling fulfilled in life. (Artaud, 1958: 25)

Recast in the language of theatre ecology, the actor's power owes its force not to performativity, but to virtuality – to the refusal of performance to reproduce the world as it is. The imperative, as Artaud considers it, is to resist all actualisation, to deform and de-create. Paradoxically, this insistence on remaining theatrical exposes spectators to drives and energies that can never be fully consumed, the passage to a 'non-act'.[50] What matters, for Artaud, is intensity, to release the potential for becomings as opposed to representing the outcome or object of that becoming. Always with Artaud, the gesture triumphs over the act, genesis over completion. Inevitably – and this is why Artaud appeals to Guattari and Lyotard – this exposure to an intensity without release, to *pure* affect, is cruelty in motion. Through it, the human is decentred, ejected from its position of mastery and control. The result is anguished knowledge, a heightened sense of vulnerability, the awareness that, as Artaud cautions, 'we are not free [...] and the theatre has been created to teach us that first of all' (Artaud, 1958: 79).

History   171

The alchemical and animated theatre that Artaud proposes, then, is one that is aware of the (im)possibility of ecology, the anguish involved in possessing what Lyotard calls an *oikos*. To access that *oikos*, Artaud uses theatricality to expose audiences to the wound of existence, a laceration in which 'everything that acts is a cruelty' (Artaud, 1958: 85). And yet, as Artaud also makes clear, there is little hope of escaping a world that 'is committing suicide without noticing it', if this existential distress remains untreated (Artaud, 1958: 32). To repress the suffering of the *oikos*, to remain prey to the rationalising discourses of economics and strategy, is to persevere, disastrously, with the malicious magic – 'the bewitchment' – spun by a God of Judgement (1958: 31–2). In order to evade this 'spell', Artaud's theatre ecology is deliberately oxymoronic and paradoxical. In it, weakness is a strength, creation the product of a refusal to perform (Artaud, 1958: 24). The ambition is not to mobilise but to derealise, to give a chance to the possible without offering any definitive schema or plan of action as guide to its construction.

Against an activist culture that fetishises results and optimal performance, Artaud urges theatre to serve its own ends, to exist as an event that provokes transference, that discloses, in Lyotard's terms, the trauma of a terrestrial birth that one has never truly left. In his commitment to theatre's singularity, Artaud abuts on the alternative activism inherent in the work of Pignarrre and Stengers. For the latter, the 'struggle against capitalism' has little or nothing to do with 'sober' thinking that would only be concerned with obtaining practical results. Such logic is part of the problem. On the contrary, Pignarre and Stengers encourage activists to develop an 'appetite for new types of risk', to get a taste for drunkenness, a new poetry for transformation (Pignarre and Stengers, 2011: 66–7).

There is nothing sober about Artaud's theatre, no desire on the part of 'Artaud le Mômo', the self-styled kid, to 'rationalise' the future or measure risk.[51] Almost a century before Pignarre and Stengers, Artaud knows that strategic, militaristic ways of thinking only perpetuate the 'consciousness-volition-intentionality' schema that capitalist sorcery depends upon. To provoke a revolution in what Guattari names 'mental ecology', something else is required, the fabulation of a spacetime where forgotten memories are recalled and drunkenness affirmed, an awaking of enthusiasm.

In place of capitalist rituals that confirm the status quo and approach the earth as a commodity, Artaud posits that most paradoxical of things: a ritual theatre in which nothing is produced only affects and percepts. By animating the world virtually, Artaud seeks to give birth to a collective body running on 'electric nerves', a people willing to live through their senses and in their bodies, to become other than what they are (Artaud, 1976: 517):

> Everyone down to the last coal-seller must realize
> That we have had enough of decency –
> Physical as well as physiological
> And DESIRE a fundamental
> > Change
> Of THE BODY.
>
> (Artaud, 1976: 579–80; original cases)

Like Guattari in *The Three Ecologies,* Artaud knows that the collective stakes of his corporeal project to 'remake humanity's anatomy' could not be higher (Artaud, 1976: 570). In cruel theatre, ecology is a matter of life and death, not just for humans but for planetary life, in all its forms. Haunted by what he calls 'the apocalyptic war to come', a war that far exceeds 'the conflict between America and Russia', human existence needs to be rethought and remade, Artaud proclaims. It has to be dis-*organ*ised, made terrestrial, returned to its mineral and vegetal roots, lived as an *oikology*, a new terrestrial *habitus*. As he communicated to the curator and art dealer Pierre Loeb in the spring of 1947, one year before his death:

> The time when man was a tree without organs or function,
> But possessed of a will,
> And a tree will which walks
> Will return.
> It has been, and it will return.
> For the great lie has been to make man an organism.
>
> (Artaud, 1976: 515)

At its most fundamental, and he never wavered on this, Artaud's cruel theatre is an earthly theatre: an attempt to move beyond the human in order to reconnect with ecological becomings, to walk as a tree, to move like a snake, to reach the virtual plane where everything is possible because everything is promiscuous, vibrant and alive. Artaud's enthusiasm for theatre must not be written off as the magical thinking of an idealist or as the delusions of a schizophrenic. To use Guattari's vocabulary, Artaud believes in transversality, in the notion that a shift in one domain can leak out and impact on all others, thus provoking an 'appetite for life' that would contest capitalism's total war against the earth (Artaud, 1958: 102).[52]

Conclusion

In this chapter, I have traced an alternative ecological history of contemporary theatre and performance by focusing, initially, on Michel Corvin's dramaturgical understanding of abhumanism. However, while Corvin's concept is a useful category for describing modern theatre's attack on the humanism that has been at work in theatre since Sophocles and Aristotle, it is too philosophically contradictory to stand as a valid theatre ecology. For a more progressive theatre ecology to emerge, it was necessary to turn to the theories of Artaud and to ecologise his thinking in such a way that extant critiques of his work are turned on their heads.

In the twenty-first century, it is not enough to remain at the level of critique when it comes to Artaud, even if some of that critique is merited. And neither is it useful to immunise readers against Artaud's ecology by approaching it as western metaphysics in reverse, as Jacques Derrida does. In an age of environmental exhaustion, something more positive is demanded, a recalibration that permits Artaud's theatre to be analysed as an 'ecological event', the aim of which is to release spectators from the nihilistic spell of capitalist sorcery in the name of a new earth and a planetary people. In the next chapter, I remain faithful to my 'idea' of a theatre ecology by rethinking what the style and function of performance analysis needs to become once theatricality is foregrounded as an ecological event. In accordance with Artaud's own practice, the intention is express a new type of writing by opening the body up to the animating 'touch' of cruel affects – to the virtuality of the theatrical, in other words.

Notes

1 This need to argue for the significance of Artaud's ideas is due, in large part, to the elliptical and often gnomic quality of his poetic style.
2 Though it has been touched on, the potential of Artaud's ecology has yet to be fully acknowledged by theatre critics. Reflecting on *The Ecocide Project*, her collaboration with playwright Shonni Enelow, Una Chaudhuri dedicates a single short – but crucial – paragraph to Artaud. For Chaudhuri, in an age of climate change, Artaud's existential notion of cruelty is now figured as 'eco-cruelty', a negative force highlighting 'human helplessness in the face of the transformations of nature' (Chaudhuri, 2013: 27). Baz Kershaw's engagement is even shorter. In an essay on metabolism and energy in theatre, Kershaw problematises Artaud's celebrated image of an actor signalling through the flames, interpreting it as irresponsibly wasteful: 'There is certainly energy out of place in Artaud's frighteningly memorable image. But is it ecologically responsible or irresponsible energy? And am I the first to wonder, somewhat ridiculously, if this figure draws attention to itself at the price of the conflagration's environmental effects, the impact on its place of enactment?' (Kershaw, 2007: 292). Chaudhuri's and Kershaw's respective comments on Artaud are missed opportunites. They do not explain the centrality that Artaud's work has for theatre ecology, and, in Kershaw's case, they mistake its contribution entirely by investing in a largely

theological notion of expenditure. What Kershaw fails to see is that Artaudian expenditure contests normative understandings of energy storage and wastage; it is more accurately defined as an anti-expenditure, a negentropy that has nothing in common with capitalist ideas on consumption. To get a better insight into the environmentalism of Artaudian energetics, see Allan Stoekl's reading of Georges Bataille's solar economy in *Bataille's Peak: Energy, Religion, and Postsustainability* (Stoekel, 2007).

3 It is interesting to note that Corvin's understanding of posthumanism has nothing in common with how I use it in this book. Indeed, Corvin prefers to define it, anachronistically and religiously, as 'the glorious body of the resurrection' ('le corps glorieux de la résurrection') (Corvin, 2014: 28; my translation).

4 In a note, Corvin explains that, unlike Audiberti who used abhumanism to limit the so-called exceptionalism of the human, he uses it to highlight its protean power, the capacity of the human to be everywhere and everything (Corvin, 2014: 18). I will come back to the lack of ecological sensibility in Corvin's understanding of abhumanism in the main body of the text (see pp. 143–7 in this chapter).

5 One of the oddities of Corvin's book is that he shows no interest in sociology or history in his study of abhumanist theatre. Like many classic avant-garde studies in Theatre and Performance Studies, the aim is to focus on representation alone. While there are possibilities opened up by such a method, it also tends to autonomise the artwork in a reductive way, to void it of any kind of social and political context. These 'gaps' in his study are what my focus on Artaud's ecology is intended to counter. An alternative approach would be to think through the economic, scientific, militaristic and technological developments behind the different types of abhumanist practices that Corvin is concerned with.

6 In humanist theatre, as understood by Corvin, the actor hosts the character in such a way that no gap or distance exists between them, a suturing that is repeated by the spectator who takes this simulated figure as its *semblable*, its double. The friability, perhaps even futility, of this dramaturgical project is disclosed when one recalls that the very being of theatre militates against such a total act of identification. To watch theatre, as Plato knew, is to be aware that the actor both is *and* is not the character they play. There is always an irreparable oscillation or syncopation when a human appears on a stage, a movement that discloses the emptiness on which theatre constructs its fictions. This vertiginous and constantly moving borderline – what Corvin refers to as an 'interval' (*'écart'*) (Corvin, 2014: 18) – undoes the humanist project theatrically and philosophically. *Theatrically*, because it highlights the artifice of the character; *philosophically*, because it troubles the idea that the human being has a substantive essence or proper ground for its being.

7 Corvin says that abhumanist theatre shows the human to be 'supra-human (the divine), ante-human (materiality), in-human (the robot), post-human (the "glorious" body of the resurrection), para-human (the machine), circum-human (the past, ghosts), anti-human (death), infra-human (the animal)' ('supra-humain (le divin), l'anté humain (la matière), l'in-humain (le robot), post-humain (le

corps "glorieux" de la resurrection), para-humain (la machine), circum-humain (le passé, les fantômes), l'anti-humain (la mort), l'infra-humain (l'animal)'). (Corvin, 2014: 28; my translation).

8 In ecological terms, it is not just the human that is '*en trop*', as Corvin suggests; it is life itself, with its indeterminate process of individuation insisting on new possibilities for existing, demanding an infinity of performances to come.

9 This theory of theatrical individuation is very different from how a humanist thinker like Jonas Barish understands it. For him, individuation is psychological and existential, something that only applies to human subjects and which results in a movement towards authenticity and self-realisation. For more on theatricality as a mode of self-realisation, see Worthen (2020: 26–31).

10 Earlier in his study, Corvin rightly insists that it is not enough to challenge the anthropocentrism of the stage by merely representing animals and gods in the manner of Classical drama. Something else is required: the production of actants that would *really* stand apart from humans: androids, machines and puppets. However, as I explain in the main text above, Corvin is reluctant to see that this abhumanist mode of appearing is still coterminous with a humanist desire for full presence. Anthropomorphism is more difficult to dislodge than anthropocentrism. For where anthropocentrism can be challenged by representing actual non-humans on the stage, anthropomorphism is a projective activity that needs to be tackled in a different manner, via a different practice of signification. As I show in my reading of Artaud, such a strategy is predicated on affect rather than representation.

11 Craig states that he: '[P]rays earnestly for the return of the image … the Über-Marionette, to the Theatre; and when he comes again and is but seen, he will be loved so well that once more it will be possible for the people to return to their ancient joy in ceremonies … once more will Creation be celebrated … homage rendered to existence … and divine and happy intercession made to Death' (Craig, 1908: 15; original punctuation).

12 Von Kleist notes how 'grace returns after knowledge has gone through the world of the infinite, in that it appears to best advantage in that human bodily structure that has no consciousness at all – or has infinite consciousness – that is, in the mechanical puppet, or in the God' (von Kleist, 1972: 26).

13 This reading of Artaud's cruel ecology differs a little in emphasis but not in substance from Jay Murphy's brilliant text *Artaud's Metamorphosis: From Hieroglyphs to Bodies without Organs* (2016). For Murphy, Artaud moves from an obsession with finding the missing hieroglyph 'as a kind of esoteric or original language' towards a more anarchic, affect-driven project founded on the production of the 'body without organs', a plane or plateau of existence which erases all origins, signs, and lines (Murphy, 2016: 2). While I certainly agree that there is a distinct move away from theatrical signs and gestures towards voices and bodies in Artaud's work, I am perhaps more inclined than Murphy to read for that virtual, pre-individual force in his theory of cruelty in the 1920s and 1930s, too.

14 It is for this very reason that I have such admiration for the 'realism' inherent in Christel Stalpaert's response to Kris Verdonck's work. Acknowledging her own

distress in the presence of Verdonck's performances, she says: 'It is not from within the illusory safe references points of chronological time but from within a mode of uncertainty that an openness towards the to-come emerges' (Stalpaert, 2020: 130). In this way, Stalpaert recognises the need for an ecology of rupture.

15 Artaud is also more of a materialist. Corvin's abhumanism falls into the Kantian trap of correlationism, the contention that knowledge is idealist, unable to get beyond the structures that humans impose on the world.
16 For key readings of cruelty in Artaud, see Martin Esslin (1976), Pierre Brunel (1982), Monique Borie (1989), Jane Goodall (1990, 1994), Stephen Barber (1993) and Helga Finter (1997).
17 Artaud's 1946 drawing the *Théâtre de la cruauté* shows a number of female bodies – what he refers to as his 'daughters' – encased in coffins and in proximity to a nebulous form, a distorted body that reminds one of the figures in Francis Bacon's paintings.
18 The 1934 plays *The Conquest of Mexico* and *Heliogabalus or The Crowned Anarchist* form part of the Theatre of Cruelty project, too.
19 Artaud gets close to the ecology that interests me, when he says that Van Gogh's painting allows something in 'Nature' to discover itself anew and to become other than what it is. For Artaud, art already is 'Nature', an expression of life's abundant creativity, something processual (Artaud, 1976: 484, 489).
20 'There is nothing inert, nothing benevolent in Gaia' (Latour, 2017: 108).
21 I take the word 'irreparable' from Tsu-Chung Su's fascinating reading of Artaud's account of his journey in Mexico in 1936–37. Su borrows the term from Giorgio Agamben for whom it signifies 'things just as they are […] without remedy to their way of being' (Agamben in Su, 2012: 8).
22 Gnosticism is a negative religion that sees the earth and material existence as primordial errors, mistakes – the work of a demiurge. For the Gnostic, the transcendent being is immaterial, all spirit, nothing grounds it. As with Platonism, proper substance is an idea in Gnostic philosophy. It can never be realised in the world. As Goodall puts it, Gnostic literature describes earthly existence 'as numbness, sleep, drunkenness or oblivion' – something to exit, not become with (Goodall, 1994: 75–6). Such a reading seems curiously inconsistent with Artaud's passionate attachment to the earth, and also to his desire to bring body and mind together in 'a spiritual physiognomy' (Artaud, 1958: 22).
23 A very different, non-Fascistic and pro-feminist reading of Artaud is made by the artist Nancy Spero in her series *Codex Artaud* (1971–72). For more on Artaud and feminist performance, see Maryann de Julio's essay 'Nancy Spero's *Codex Artaud*' (1997), which focuses on Spero's relation with Artaud, and, more recently, Lucy Bradnock's monograph *No More Masterpieces: Modern Art after Artaud* (2021) that traces Artaud's influence on the US avant-garde, and, in particular, on the thinking of Spero and Carolee Schneemann.
24 In this context, see Constance Spreen's excellent account of the Artaudian plague as a reversal of the pathologisation of disease by Charles Maurras, the leader of French proto-Fascist group Action Française. Spreen's reading is in direct contrast with Jannarone's. Spreen notes how 'against the plagues of

violence and excess in all forms that afflicted modern life, especially nationalistic fervor, the theater as plague was a badly needed form of resistance' (Spreen, 2003: 96).

25 In her 2008 collection *When Species Meet*, Haraway explains that, for Margulis, 'everything interesting on earth happened among the bacteria, and all the rest is just elaboration [...]. Bacteria pass genes back and forth all the time and do not resolve into well-bounded species' (Haraway, 2008: 31). The vitalism and vivacity of this process leads Haraway to argue for the necessity of affirming contamination and contagion. For her, life is 'a multi-partner mud dance issuing from and in entangled species [...]. These are the contagions and infections that wound the primary narcissism of those who still dream of human exceptionalism' (Haraway, 2008: 32).

26 Borrowing from earlier critiques made by Susan Sontag (1976) and Rustom Bharucha (1978), Jannarone's argument is predicated on the fact that Artaud, apparently, showed no interest in the cultures he appropriated, preferring to hallucinate them into being, and conflating them as an amorphous mass of alterity. In Jannarone's view, he lacked, then, a properly ethnographic ethic for engaging with Indigenous peoples and based his knowledge claims on personal and orientalist fantasy of the 'noble savage'. However, while it does not excuse Artaud's neo-colonial tendency to speak on behalf of the Tarhaumara Indians or indeed to fantasise about them, Raymonde Carasco (2006) has produced an exhaustive account of how Artaud's description of the Tutuguri rituals in Northern Mexicio shows evidence of profound ethnographic sensitivity and cultural knowledge. Similarly, Jay Murphy (2016) has traced the longstanding effect that Artaud's journey to Mexico had on his life and work. Artaud did not simply appropriate Tarahumara culture in order to remain the same. He was changed by his engagement with it and sought to understand it over a long period of time. Such an approach is not, in any way, the act of a 'classical orientalist', someone who, in Edward Said's influential reformulation of the term (1978), is concerned to keep the 'other' at a distance and is terrified of the very thing that Jannarone equates with Artaud's Fascism: contamination and contagion. Equally, Nicola Savarese is aware that Artaud's reading of Balinese theatre, though orientalising in its failure to critique the colonial context in which it was performed and distorting in its ability to make claims about a culture he knew nothing about, is nevertheless politically and aesthetically productive in the extent to which he is sensitive to the skill and technique involved in Balinese theatre's alternative mode of relating to and representing the world. According to Savarese, by going beyond surface exoticism in his response to Balinese theatre, Artaud not only intuited its 'spirituality' (Savarese, 2001: 71), he blasted it out of an equally dangerous, primitivist fantasy, the one that would seek to guard its purity, keeping it safe from the 'alchemy of cultural processes' (Savarese, 2001: 77). An alternative method for rethinking Artaud's engagement with non-western practices and ideas is formulated by Eve Kosofsky Sedgwick when, in her writing on Buddhism, she advances the idea of recognition/realisation as something different from appropriation and adaptation (Sedgwick, 2003: 156).

27 Clifford cites Artaud several times in *The Predicament of Culture* as an exponent of ethnographic surrealism and subject of displacement (Clifford, 1988: 127, 129, 152).
28 Others critics who are, perhaps, too quick to accuse Artaud of primitivism, Orientalism and cultural appropriation, include Christopher Innes (1993), Julie Stone Peters (2002), Dietrich Harth (2004) and Min Tian (2018). Monique Borie (1989) offers a more balanced account.
29 According to Lewis and Maslin, the Columbian exchange marks the early modern moment when the 'cross-continental movement of species, plants, peoples and diseases between Europe and Americas 'contributed to a swift, ongoing, radical reorganization of life on Earth without geological precedent' (Lewis and Maslin, 2015: 174).
30 Jannarone points out that she is providing a 'historical approach to Artaud that has been avoided up to this point' (Jannarone, 2010: 25). However, nowhere does she discuss Artaud's attack on US imperialism and neither does she consider that Artaud's aesthetic and political mindset is utterly opposed to western capitalist modernity – the very thing that Fascism sought to restore.
31 In this context, there is also much in Artaud that resonates with what the anthropologists Danowski and Viveiros de Castro call *'becoming* and *rebecoming-indigenous'* – the desire to move away from western models of eschatology for a more generative ecological approach to time found in the narratives of Indigenous peoples (Danowski and Viveiros de Castro, 2017: 122; original italics).
32 One has only to look at the technology involved in *Pour en finir avec le jugement de Dieu* to see that Artaud is not interested in any kind of authentic transposition of Tarahumura rituals.
33 See here James Clifford's latest text *Returns: Becoming Indigenous in the Twenty-First Century* (2013), the final part of his critical ethnographic trilogy that started in the late 1980s.
34 Jannarone and Bharucha tend to downplay the positive politics in Artaud, leaving readers with nothing in reserve to combat the current status quo, no tools to fight capitalist sorcery. While I am always aware of the dangers in Artaud's cultural politics, I prefer to adopt a more affirmative stance, taking what I can for the purposes of the present. This position is what I mean when I talk of creating alliances, not identifications, tactics of ecopolitical alignment.
35 Derrida's critique provides, in many ways, the philosophical ground for Goodall's and Jannarone's readings. The 'turn to theory' within Performance Studies that started in the 1980s has been marked by a suspicion of unmediated experience, something that western practitioners in the 1950s and 1960s were thought to be drawn to, and which Artaud was seen to be one of the primary instigators of. Despite some exceptions, Derrida's reading of Artaud has remained largely unchallenged by theatre scholars, an example of a philosopher cautioning against naive understandings of performance.
36 Not only is the title a pun on the late text *Artaud le Mômo* (1947), but Derrida's essay is a *détournement* in the Situationist usage of the word. For Derrida's

lecture, delivered in a museum, is an attack on the museumification of Artaud. He ends the lecture at MoMA with an allusion to Artaud's incarceration in Rodez, thereby suggesting yet another violence committed on Artaud, even after his death: 'Poor Artaud. What is happening to him! He will have been spared nothing, this Mômo. Nothing. Not even the survival of his spectre, not even the most equivocal, and cruelly ambiguous, the most vain and most anarchronistic revenge' (Derrida, 2017: 77).

37 While late Derrida deals with subject matter that exists in proximity to environmental issues – see, for instance, his writing on nuclear disarmament in '"No Apocalypse, Not Now" (Full Speed Ahead, Seven Missiles, Seven Missiles)' (Derrida, 1984) and work on animals and industrial farming in texts such as *The Animal that Therefore I Am* (Derrida, 2008) and *The Beast and the Sovereign, Volume 1* (Derrida, 2009) – he is nevertheless silent about ecology. Jonathan Crary's comments in *Scorched Earth: Beyond the Digital Age to a Post-Capitalist World* (Crary, 2022) are telling, in this regard. Speaking of Derrida's 'ten plagues of new world liberalism' as listed in *Specters of Marx: The State of the Debt, the Work of Mourning and the New International* (Derrida, 1994), Crary is quick to point out that there is 'not a hint of impending ecological catastrophe or of capitalism's contribution to mass extinctions and the collapse of ecosystems' (Crary, 2022: 34).

38 To use the terminology of his later writings that touch indirectly on ecological matters, Artaud would doubtless be accused by Derrida of adopting a mistaken attitude to 'biodegradability'. On one hand, the biodegradable object is something to champion, for Derrida, since it has otherness built into it from the very beginning. And yet, on the other hand and at the same time, it is something to guard against since it threatens to disappear completely, to assimilate itself into the earth, and thus complete an operation that leaves no trace, no possibility of an afterlife, a celebration of oneness. For more on this, see the essays by Michael Naas (2018) and Michael Peterson (2018) in the collection *Eco-Deconstruction: Derrida and Environmental Philosophy*.

39 In French, the second syllable of the word *inné* is 'né', the past participle of the verb *naître*, which translates as to give birth, to be born. There is, then, a sense of 'birthing' at work in the innate, a process of becoming.

40 Deleuze and Guattari explain that: 'There is a mode of individuation very different from that of a person, subject, thing, or substance. We reserve the name haecceity for it. A season, a winter, a summer, an hour, a date have a perfect individuality lacking nothing, even though this individuality is very different from that of a thing or subject. They are haecceities in the sense that they consist entirely of relations of movement and rest between molecules or particles, capacities to affect and be affected' (Deleuze and Guattari, 1987: 261).

41 It is interesting to compare Derrida's deconstruction of Artaud with his critique of Rousseau's festive performance in the essay 'That Dangerous Supplement' in the book *Of Grammatology* (Derrida, 1974). In both instances, anti-theatrical performance is proposed as a way of getting back to 'Nature', a vehicle for self-presence, according to Derrida. However, a distinction needs to be made.

While Rousseau rejects theatricality, Artaud embraces it and looks for alternative modes of representing within theatre itself.

42 Derrida maintains that Artaud was aware of the impossibility of his project, and that the limit he 'kept himself as close as possible to' is a victory in spite of itself (Derrida, 1978: 314). Artaud's tragedy – or 'naivety' – was to continue to transgress this limit, to refuse it (Derrida, 1978: 244).

43 As Deleuze and Guattari put it on their gloss on Artaud's concept: 'You never reach the Body without Organs, you can't reach it, you are forever attaining it, it is a limit' (Deleuze and Guattari, 1987: 150).

44 This process refers to the transformation of lead into gold, in the case of the alchemist; and body to idea, in the case of the theatre artist.

45 Hylomorphism propounds a dualistic ontology in which matter (*hylē*) does not fulfil its potential until an external form (*morphē*) has been imposed upon it. Artaud contends, by contrast, that form is already at play within matter.

46 Stengers explains how animism functions as a performative 'intercessor' by drawing on Deleuze and Guattari's notion of the milieu, a territory or context in which meaning and subjectivities are produced and take on a consistency: 'I once offered the example of the Virgin Mary – not the theological figure but the intercessor that pilgrims address. It's wrong to think the Virgin Mary could make her existence known independently of the faith and trust of pilgrims; for her to do so in a situation committed to the question of how to represent her would be in bad taste. Rather, if we accept that the aim of the pilgrimage is the transformative experience of the pilgrim, we must not require that the Virgin Mary "demonstrate" her existence to prove that she is not merely a "fiction". [...] Instead we must conclude that the Virgin Mary requires a milieu that does not answer to scientific demands' (Stengers, 2012: 3). I would suggest that Artaud's animist notion of theatre be read in a similar way, especially with respect to Derrida's metaphysical critique of presence in his work.

47 For more on animism and western performance, see Mischa Twitchin and Carl Lavery's special edition of *Performance Research* 'On Animism' (2019), and for more on the Lisbon earthquake, see Nigel Clark's brilliant analysis in *Inhuman Nature: Sociable Life on a Dynamic Planet* (Clark, 2011: 81–106).

48 While she has great sympathy with his project, Stengers is keen to differentiate her position on animism from David Abram's. Where Abram actively enlists Indigenous modes of animist thought to undermine the West's disenchantment of the world, Stengers wants, always, to stay 'on my [the western] side of the divide' (Stengers, 2012: 1), to decolonise western thinking from the inside, to show, as Latour is also keen to, that 'we have never been modern', never wholly disenchanted (Stengers, 2012: 2). Artaud's animism is of this order, too. For him, Balinese theatre and Tarahumara rituals were always meant to be a stimulus, not a model to imitate or transport.

49 Barad comes uncannily close to Artaud's animism when she contends that 'touch is the primary concern of physics' (Barad, 2012: 208). For her, touch must be studied 'in its physicality, its virtuality, its affectivity, its e-motion-ality' (Barad, 2012: 208–9).

50 In Lacanian psychoanalysis, the 'passage to the act' refers to an event – a 'doing' – that takes place in the Real as opposed to the Symbolic. It is often associated with trauma and psychosis, an acting out. The passage to a 'non-act', by contrast, is a catalyst without fulfilment, a performance that refuses to perform.
51 'Le Mômo' (French slang for 'kid') is a nickname that Artaud gave himself and used as the title for his 1947 publication, *Artaud le Mômo*. See also pp. 178–9, note 36 in this chapter.
52 Inherent to Artaud's idea of doubleness is the avant-gardist commitment to heterogeneity, the sense in which theatre and life are disparate parts of a single continuum that refuses to resolve into a whole. Or as Artaud put it in the 'Second Manifesto of Cruelty': 'Between life and theatre there will be no distinct division, but instead a continuity' (Artaud, 1958: 126).

6

Analysis (becoming archipelagic)

Introduction

Chapter 6 – the final chapter in this first volume – is intended to perform an interstitial role. As well as pointing back to theoretical, lexical and historical points I made in Chapters 3, 4 and 5, it also takes a swerve in a different direction, gesturing towards what is to come in the volumes *Theatre and Landscape: Taste, Ecology and Politics* and *Theatre as Clinic: Constellations of the Anthropocene(s)*. The intention is to demonstrate a practice (not a method) of performance analysis that would do justice to the ecological possibilities of the theatre event by encouraging critics to set sail on what Rachel Fensham evocatively terms 'the ebb and flow of a sensate event' (Fensham, 2009: 11).[1]

In order to 'tack' the ecological swell which Fensham's language is so suggestive of, this chapter follows the unfurling of a single, but discontinuous wave of thought.[2] In the first motion, I propose a new theory of ecoperformance analysis by returning to Lyotard's theory of the *gestus*, an aesthetic trope that allows one to understand how artworks bring the cruelty of the *oikos* into play; in the second, I advance the archipelago as a critical *topos* for ecologising theatre scholarship in such a way that the remit and terrain of analysis itself is changed; in the third, I provide two short examples of what an archipelagic analysis looks like. In keeping with the invitation inherent in the archipelagic thinker Édouard Glissant's notion of 'relation-relay',[3] the ecocritical focus is no longer on interpretation, but on negotiating the ecological impact of the performance event itself; in how, that is, it often places you, the spectator, at the mercy of a 'gestural force that […] [operates] from within experience itself, activating a shift in tone, a difference in quality', as the dance scholar Erin Manning puts it (Manning, 2016: 1; original citation modified). Confronted with this ecology of force, this terrestrial power, the critical imperative is to learn how to channel and express it, to fashion a set of principles for becoming archipelagic.

Analysis 183

Gestic criticism

In common with Patrice Pavis who provides the salutary reminder that 'no general mise-en-scène exists', I have little interest in fashioning an overarching theory of performance analysis that would be applied monolithically to every performance event (Pavis, 2003: 303). My ideas are entirely pragmatic and empirical, occasioned, as they are, by the cruel but singular logic of the performance in question.[4] The theatres that interest me do not deal with environmental questions in explicit ways. In them, there is no recognisable world beyond the stage to intuit, no message to impart. In fidelity to Artaud's injunctions, they are predicated on producing intense, affective experiences that leave audiences disorientated and bewildered, at the mercy of posthumanist energies they cannot control.

In these cruel interpellations and sensate diffractions, the ability to stand at a distance in order to interpret the performance event no longer holds (Van der Tuin, 2014: 239). Not only is it impossible in this contact zone to dissect the work piece-by-piece, but the call of the performance, its 'sound and fury', breaks on the body of the analyst, demanding that they find some way of accounting for the disturbance it has caused, provoking them into a response, whether they are comfortable doing so or not. Barad explains why:

> *Responsibility is not an obligation that the subject chooses but rather an incarnate relation that produces the intentionality of consciousness.* Responsibility is not a calculation to be performed. It is a relation always already integral to the world's intra-active becoming and non-becoming. It is an iterative (re)opening up to, an enabling of responsiveness [...] an on-going rupture, a cross-cutting of topological reconfiguring of the space of response-ability. (Barad, 2014: 183; original italics)

Barad's theory of responsibility holds philosophically but its aesthetic potential remains undeveloped, entirely latent. To concretise its possibility, it is necessary to unfold Lyotard's little-discussed notion of the *gestus* that I merely glossed over in Chapters 2 and 3 (see pp. 96, 113). For Lyotard, the *gestus* is not a gesture as conventionally understood, that is, a codified movement or mannerism made by a body that situated cultural agents are able to decipher and/or reperform, as anthropologists and theatre theorists such as Marcel Mauss (1973), and Eugenio Barba and Nicola Savarese (1991) have maintained. Rather, the *gestus*, as Lyotard describes it, is a 'thrust of matter' (Lyotard, 1993b: 40), an 'overflowing profusion' (Lyotard, 2009a: 89), a 'thing without a name', 'the other of the concept' (Lyotard, 2009a: 83).

Differently from Bertolt Brecht's more widespread understanding of the term (see Chapter 3, p. 19), the *gestus*, in Lyotard's view, is *not* associated with the ability to 'read' an occulted contradiction that ideology disassembles and that directors and performers are able to bring to the surface of consciousness through a self-referential mode of signifying.[5] Rather for Lyotard anything that draws attention to its own opaque and inscrutable materiality can be a *gestus*. When this 'return' to matter takes place, an 'in-significance' surges forth – a blindness, a noise – which refuses to be '*digested*' by conceptual thought (Lyotard, 1993b: 38; original italics):[6]

> As with the visible, there squats in the thinkable, a ventriloquist host, unknown, a large animal savage and mild, the child who breaks and complains. Inside language there would be a rumour that does not speak, just as inside colourful appearances there is a pale vibrato that does not see. (Lyotard, 2009a: 109)

In so far as the *gestus* thwarts 'assimilation' by the intelligence (Lyotard, 1993b: 41), it is a perfect instrument, a virtual catalyst for allowing the *oikos* to make an 'appearance' – if indeed 'appearance' is the correct word for something that cleaves so tightly to the dark, to shadow. Recalling my short exegesis in Chapter 3 (see pp. 110–13), the *oikos* eludes speech. It expresses itself as a 'rumour', a stammer of distress and abandonment that the human feels in the sheer facticity of existence, when it is touched and wounded by an unspeakable thisness, described by Lyotard as 'a suffering that is at the same time the mere condition of thinking and writing' (Lyotard, 1993a: 106).

Without ever going so far as to formulate an eco-aesthetic theory or determine a specific course of environmental action, Lyotard, in a series of oblique, fragmentary engagements in his late writings, demands that any ecology worthy of its name ought to be preoccupied with this suffering, to be willing to affirm it.[7] There is 'something' in the *oikos*, he explains, that does 'not want the bucolic', but instead is at war with the 'ruses' and 'busyness of the ego' (Lyotard, 1991: 196). The role of the *gestus* is to tap this bother, to materialise it, to transform suffering into an affirmation of life, 'an event of a passion' that is accepting of both 'anguish' and 'jubilation' (Lyotard, 1991:141).

Like the *oikos*, which it simultaneously channels and catalyses, the *gestus* 'is not only a challenge, but a denial' (Lyotard, 2009a: 45). And what it denies is absolute knowledge, the capacity of the *logos* ever to capture and control the artwork, the singularity of aesthetic experience. The question posed by the *gestus* – 'Will you ever know me?' – cannot be answered in the realm of words but only in our ability to pay attention to feeling, Lyotard says (Lyotard, 2009a: 131). For feeling, Lyotard continues, is the quality that leads to the intuition of a 'meaning that was present before being present' in some immemorial night of creation, in a memory before memory (Lyotard, 2009a: 79). If, then, as Lyotard never tires of repeating, the *oikos* returns through a kind of 'anamnesis', a memory of something that we cannot recall because it has never left us, it goes without saying that the role of eco-analysis must also change, and profoundly so.[8] The point is not to comprehend the work but to recognise a debt to it:

> Debt implies the claim of an other on the self. I have called the creditor this *thing* which is encrypted in the unconscious, not of the painter or writer, but of painting or writing [...]. Are the arts of painting or writing ethical, not for their represented motifs, nor their ends, but because they are prompted by an obligation? (Lyotard, 1999: 26)

Lyotard's question is rhetorical. The *gestus* of art carries a debt not because of what it says but because it puts us in contact with the creativity of life itself, the desire of matter to become other than what it is, to diffract and de-phase.[9] Significantly, though, the becoming that the *gestus* bestows is inseparable from a sense of suffering. Once again, we are in the realm of the *pharmakon*, the space of the poisonous cure:

> Such is cruelty. Just as language impedes writing, so the visible impedes painting. This is where there is in the work the terror of a loss suspended within the sensible. The visual work makes one feel as though one's eyes have been abandoned. It makes the blue, the red, the oblique return. It is an event, a birth, but always a melancholic one. (Lyotard, 1999: 30)

Unlike the majority of ecocritics who are content to efface themselves as bodies in their interpretations, Lyotard demands that sensations provoked by the artwork's 'sad ecology' be expressed in critical commentary (Lyotard, 2009a: 163). Not to do so is to renege on the ecological obligation that critics owe to the anamnesiac quality of the artwork, its reminder that the earth has no need of humans to speak for it. The last thing performance needs, especially ecoperformance, is a philosopher, obsessed with meaning, clarity, interpretation:

> If we philosophers are to divest ourselves of our knowledge in order to accede to the gesture that is the work and transcribe it within our own space-time-matter, we must obviously take care not to take this space, time, and matter itself for granted. Here I am referring to the space-time-matter of philosophical discourse itself [...]. Our debt to art requires that we take care of that within and toward which our own phrases arise as we try to comment on or to interpret works of art [...]. Consequently, we must rid our discourse of the referential, cognitive or objectifying function that philosophy, like other intellectual disciplines, naively and unconsciously confers upon words and their arrangement [...]. (Lyotard, 1993b: 41–2)

Lyotard's, perhaps ironic, disdain for philosophy's language of objectivity, its desire to colonise, does not result in a lack of rigour, however. Lyotard is adamant that not every response is valid. Against those critics who have sought to mimic the work or to use it as a mere catalyst for their own poetic flights of fancy, Lyotard proposes that criticism should remain (impossibly) faithful to the work; it must never lose sight of the singularity of the *gestus* that brought it into being:

> It is certainly not a question here of acquitting ourselves of our debt of transcription by miming the work of art that we are to transcribe. Moreover such an imitation makes no sense [...]. Instead, we must receive and welcome rhythms, virtual sonorities, lines, angles, curves and semantic colours, while also adroitly awakening the semantic layers which lie dormant within words and their sequences. [...] In other words, they must begin to think about thought itself as a work of art and no longer as an argument. (Lyotard, 1993b: 42–3)

Another way of putting this would be to say that thinking is compelled to be critical and creative at the same. This doubled thinking can be accomplished only by foregrounding the body as a medium of sorts, a type of surface that is attuned to the ungraspable transmission of the work:

> All that you are permitted to invoke – neither your fidelity as a witness, nor the sincerity of your homage, nor the labor you had to perform – is your body. Singularity is of the body but precisely of a body that is so singular that you do not know it or anything about it. Singularity is of this body which is not this matter, this life, and this body which are reputed to be yours by the doctor, the jurist, the sergeant recruiter, the manager and the sexologist. It is neither the sensorimotor body of the sociologist, nor the acculturated body of the anthropologist but the monster inhabited by the thing and because of this, belonging potentially to other spaces, times, matters than those of which the experts and your bodily consciousness are aware. All the bodies encrypted in your body are the indeterminate possession and non-place of the thing without a face. (Lyotard, 1999: 29)

In the context of this book, the body that the *gestus* brings into play is Artaudian, a body that belongs to no one, a monstrous body, an earthly body. To write this ecological body, it is not enough to rely on the autobiographical 'I', as so many ecocritics, including Marranca, Kershaw and Read, have had recourse to in their desire to situate themselves. Something less centred is needed. For in the realm of the *gestus*, the subject is made indiscernible, bereft of fixed qualities, caught up in what Lyotard variously terms 'the fragile fabric of shifting relations', 'pure movement on the spot', 'a push of matter' (Lyotard, 2009a: 106–7, 161, 41). Interrupted by the dynamic materialism of the *gestus*, and called into question by it, the ethical imperative is to affirm criticism as an art of incompleteness, what Lisa Robertson calls 'nilling' (Robertson, 2012) – a theatricalised mode of writing in which signs 'hollow' themselves out in the very act of their expression, disclosing, in the manner of the actor who is always at least double, 'the impenetrability of the work to thought' (Lyotard, 2009a: 41).[10] Like the hand of the Artaudian performer, the one whose gestures point beyond itself to an earth it can never hope to grasp, so critics ought to undo their own idiom, to evoke the rumble of the virtual within it, the terrestrial movement that no signifier can capture.

This 'gestic' or *oikological* approach to performance analysis differs, quite markedly, from the dominant methods of 'doing' ecocriticism in Theatre and Performance Studies and the Environmental Humanities. Gestic criticism stays with and moves from the performance. The prepositional emphasis is crucial. It discloses both a disposition and a practice, one that implicates and elaborates. To be implicated is to tell the 'truth' of a performance event, while all the time knowing that 'this *everything* that has to be said will, in fact, never be said' (Lyotard, 2009a: 89; original italics). The impossibility inherent to such an account diverges necessarily from that of the witness, the figure that has been fetishised by performance theorists and practitioners alike for so long now.[11] Whereas the witness confirms what they see, the implicated analyst, by contrast, is all too aware that the *gestus* of the work blinds them.

Prevented from witnessing, the implicated analyst is compelled to elaborate, to take the 'matter' of the performance in new directions, exploring unexpected territories, making a series of diffractive connections. Elaboration, here, resonates with the Freudian notion of *Durcharbeitung*, recalibrated aesthetically by Lyotard as an act of *perlaboration*, 'the working' through of an affective charge, a touch that innervates.[12] Artworks, for Lyotard do not represent worlds; they world and un-world at the same, allowing for transferences that bypass the intelligence, creating confusions between *gestus* and *oikos*, inside and outside (Lyotard, 2009a: 69). In this confusion, taxonomies and categories start to shift and mutate, different sympathies emerge.

Mineral being comes to the fore. Standing 'in front of paintings' in her grandmother's room in North Toronto, Lisa Robertson describes how she was able to

> feel an inner vibration. It entered flatly through the entire surface of my body I let myself go blank. [...] I came to think of the mute mineral affinity that accompanied my blankness as a psychic life of pigment. [...] This pigment didn't have anything to do with representation or style, yet it was dependent on the proportions and specificities of mixture. I think my feeling for painting is a deferred material telepathy, an elemental magnetism. I was noticing a mineral sympathy of my body's iron and copper and calcium towards paint. I learned to still myself to make room for this strange reception. In the spare

room, I first came to the recognition that I could be changed by these little documents of admixture, through the simple attention of a slowed, non-linguistic perceiving. (Robertson, 2020: 41)

Robertson's experience of looking is close to Lyotard's. For both, the ecology of the artwork is found by allowing the *gestus*'s thrust of matter to produce 'non-linguistic' individuations, changes in the very substance of one's being, in how it is constituted as a body. Félix Guattari would also concur. In his writing on performance, Guattari states that the event provokes a-signifying experiences which give birth to new existential territories for individuals and collectives alike. Tellingly, like Robertson, he compares those ecological transferences to telluric enrichments, the production not just of aesthetic forms but of 'modalities of being', 'a different metabolism of past-future' (Guattari, 1995: 90):

> In a more general way, every aesthetic decentering of points of view, every polyphonic reduction of the components of expression passes through a preliminary deconstruction of the structures and codes in use and a chaosmic plunge into the materials of sensation. Out of them a recomposition becomes possible: a recreation, an enrichment of the world (something like enriched uranium), a proliferation not just of forms but of the modalities of being. (Guattari, 1995: 90)

Using Guattari as a collaborator, the one who, as I said in Chapter 3, joins everything up, it is easy to see how the intense affectivity of the *gestus* performs 'transversally', establishing a line of connectivity that enfolds subjectivity into the three ecologies of mind, politics and 'Nature'. These alliances proliferate, for Guattari, because the *gestus* is plugged into pre-individual energy flows that glue or cathect the subject to the multiple planes of being they are part of: social collectives, landscapes, skies, more than human things. For Guattari, every 'I' is already a 'We', a multiplicity immersed in a logic of conjunction rather than mere connection, an earthly being, a terrestrial body.[13]

Reiterating a point from Simondon (see pp. 97–100), to write the ecological body of performance – the one that the *gestus* provokes – is not simply to express affects that have been understood and named as 'emotions' or 'intelligences' or to return the work to 'established' meanings and contexts. It is rather to catch the trajectory of a process of individuation as it *is* happening, a de-phasing that, in its rupturing of identity, calls out for what I have been arguing for throughout this book: a new earth and a planetary people. The *gestus* instigates a process. By thrusting itself beyond the work like a projectile, it asks spectators to meet the performance half-way, in an oscillating and materialised non-place, on a *plateau* that transmits magnetic waves and electric currents. Like a Zen garden, theatre is elementally promiscuous; it can transform rock to water, sand to sea, bodies to stone. It opens out, virtualises and drifts.

Figuring analysis

In keeping with Lyotard's ethics of writing, the ecoperformance analysis that I am advancing 'knows' from the very outset that not every base can be covered. Something – an excess, an abundance – always escapes the analyst. Any account proposed will be contingent and partial, a mere flake of the performance's totality. In its encounter with bodies, speeds, rhythms and undulations, the only law that ecoperformance analysis adheres to is a law of the virtual. This is an impossible law, a partial law, undercut with a void and mixed 'up with the thing relayed as well as the thing related' (Glissant, 1997: 18).

Reflecting on the gait of a 'ghostly young man' as he makes his way through the hills overlooking the Black Beach at Le Diamant on the South Coast of Martinique, Glissant makes a telling remark. 'It doesn't feel right to have to represent someone so rigorously adrift', he says 'so I won't describe him. What I would like to show is the nature of his speechlessness' (Glissant, 1997: 122). The same reserve ought to hold for the theatre ecologist, too. Confronted with the swell of the performance, the point is not merely to describe but to express the 'nature' of what cannot be said.

In the attempt to evoke this speechlessness, this gestic excess, my approach draws on the practices of theorists, such as Peggy Phelan (1993), David Williams (2006), Sarah Wood (2014), Kélina Gotman (2015, 2018), Joe Kelleher (2015a), Augusto Corrieri (2016) and Katja Hilevaara and Emily Orley (2018), all of whom look to perform their ideas by disrupting the conventions of criticism itself.[14] That not all of these writers are self-defined ecocritics is beside the point. What matters is how they produce creative readings that articulate a sense of being undone and opened up, a becoming intimate with things. Instead of explaining or mastering a work from a distance, they look to get messed up in it, to remain in contact, somehow, with Artaud's cruel genesis, the event that sets in motion the 'play' of the pre-individual.

At a stylistic level, the craning of these critics towards an 'expansive horizon' (Gotman, 2018: 8), explains my own fondness for infinite verbs. An attempt, on my part, to reach out to something vast, to leap beyond language *through* language. It also explains the importance I attach to breath, cadence and rhythm in my writing, those affects that shape experience on the far side of thought, the side beyond the intelligence, the side that is speechless. By making this commitment, performance analysis should, if it can, become a song, a way of capturing some of the noise that Artaud heard in theatre's capacity to construct 'waves of matter' crushing into each other. But how to translate this 'noise', to make ecological sense of it, in such a way that ecoperformance analysis can be applied to all forms of theatre and performance, not just works that deal with obvious environmental themes?

Perhaps by constructing a figure that would hold these disparate affects, forces and thoughts together, a poetic trope allowing the analyst to respond to the performance event as an expression of the earth's own desire for experimentation, holding together 'Nature' and culture in the fold of what I have been calling theatre ecology. In a dense, but beautiful footnote to her essay 'Time in the Codex' (2012), Robertson explains that the figure exceeds itself, becoming more than metaphor. It has, she points out, 'an agency', a capacity to 'modulate perception' (Robertson, 2012: 12). This modulation is exactly what I want my figure to provoke, to use analysis to express the transformative potential of performance, its agentic power.

In the spacetime of the figure, there is no need for theatre to 'evidence' its eco-efficacy through a logic of measurement and proof, either by confecting surveys or by tracing some link between performative cause and environmental effect. Rather, the proof of efficacy resides within the *style* of the performance analysis undertaken, in the type of writing and thinking it commands, in how it expresses its belief in the earth at a verbal level. In this way, performance analysis posits itself as a creative-critical document, a textual elaboration of the work's 'virtual ecology' – a transcription of the *oikos* through the *gestus*.

Archipelagic spectating

To elaborate that ecology critically, I want to approach theatre, dramaturgically and spectatorially, as an archipelago, a topographical term which, in its literal application, refers to how a group of islands, formed either from volcanic eruptions or the movement of tectonic plates, have broken away from the continental landmass to form a fragmented, scattered geometry of lakes and rivers, but mostly seas:

> The word 'archipelago' comes from the medieval Italian word *archi*, meaning chief or principal, and the Greek word *pelagus*, meaning gulf, pool, or pond.

> Most archipelagos are formed when volcanoes erupt from the ocean floor; these are called oceanic islands. The islands of the Hawaiian archipelago, for example, were formed by a series of volcanic eruptions that began more than 80 million years ago and are still active today. (Oceanservice.noaa.gov., 2024)

This definition from the National Ocean Service website usefully describes the geomorphology of archipelagoes, but it does not say enough for my purposes. To gain some sense of how the archipelago functions in aesthetic terms, as a figure of speech, it is important to look elsewhere. In two superb publications that build on Glissant's poetic history of the islands of the Caribbean, the Australian ethnographer Paul Carter posits the archipelago as a poetico-ecological figure for leaving dry land behind. Where grounded thought, Carter suggests, is 'polderised' and 'dammed up' (Carter, 2011: 16–22), archipelagic thought is folded into the elemental stuff it engages with. In his book *Decolonizing Governance: Archipelagic Thinking* (2019), he cites an oddly monist moment from René Descartes to prove his point:

> Nature also teaches me, by these sensations of pain, hunger, thirst and so on, that I am not merely present in my body as a sailor is present in a ship, but that I am closely joined and, as it were, intermingled with it, so that I and body form a unit. (Descartes in Carter, 2019: 12)

Because they have been affected by earthly forces, the 'archipelagist' speaks in what Bronislaw Szerszynski calls the 'middle voice' – a voice that belongs to no one (Szerszynski, 2018: 139–40). To which Carter adds, 'in contrast with continental land masses, and their conceptualisations of the world, archipelagoes shape knowledge differently' (Carter, 2019: 12). They do so by allowing thought to be informed by 'elemental traces of passage'. According to Carter, '[t]he motion of wind and water [...] are the glue of the archipelago, the form of the *inter-esse*, the in-between whose interests secures the archipelagic economy' (Carter, 2011: 18).

There is, then, a different economy of writing and expression demanded by the archipelago, one that is open to possibility, to energy, to the unrepresentable, to the theatricality of ecology. For to sail an archipelago, Carter reminds us, is to undergo a 'parallax effect', to see island after island, approaching from different sides, under different skies, onboard a different vessel. Everything in the archipelago is moving, shimmering, changing shape. Sometimes, even, 'a rock can become a canoe' (Carter, 2019: 33). In archipelagoes, significance is always to be re-created, again and again.

The populations on the 'diffracted' island of archipelagoes are equally incomplete and protean (Glissant, 1997: 33). They host a bastard people – sailors, traders, refugees, the displaced, the stolen, *maroons*, revolutionaries – all of whom are opposed to the plough, the instrument of *terra firma*. In their mouths, language becomes poetic, metaphorical, creolised. In opposition to colonisers who want to impose monologues and proper speech, archipelagic populations, by contrast, insist on heteronymy, in being 'cut together apart', in speaking pidgin dialects. Drawing from his own experience of living on Martinique, Glissant confirms that the '*tout-monde*' is no universal globe whose fragments can be pieced together: it is always individuating *and* breaking apart, a system of temporary alliances in constant reinvention. Echoing Simondon, Glissant states:[15]

> Relation, as we have emphasized, does not depend upon prime elements that are separable or reducible. If this were true, it would itself be reduced to some mechanics capable of being taken apart or reproduced. It does not precede itself in its actions and presupposes no a priori. It is the boundless effort of the world: to become realized in its totality, that is, to evade rest. (Glissant, 1997: 172)

There is an important politics to the 'poeticised archipelagics' of Glissant and Carter, an attempt to decolonise governance, to wager on a planetary way of being that resonates with the attempts made by contemporary postcolonial thinkers, such as Gayatri Chakravorty Spivak (2014), Amy J. Elias and Christian Moraru (2015) and Eduardo Viveiros de Castro (2016), to move beyond western notions of the global.[16] While I, too, am concerned with such a politics, with figuring the earth as a fragmented archipelago not a closed sphere, I delay the particulars of this engagement for volumes 2 and 3 of this project. For the remainder of this chapter, I want to advance a more oblique politics, one that is found in advancing a new notion of archipelagic spectating,

To do that, I depart from the methods of scholars such as Jazmin Badong Llana (2018) and Paul Rae (2019), both of whom have written brilliantly and innovatively about aesthetic and cultural performances in the actual archipelagoes of the Philippines and Indonesia. Rather like Deleuze and Guattari who suggest that thinking can be, amongst other things, arboreal, rhizomatic and cosmic, I want to claim that theatre thinking is archipelagic because at some level of high abstraction, on some plane of consistency, something of the archipelago is found within it. Though there is an undoubted metaphorical quality to this relation, something more than metaphor is at stake. In place of the always protective carapace of metaphor, then, I propose the exposed body of the isomorph, a mathematical structure in which, as Manuel DeLanda has pointed out in his discussion of geology in *A Thousand Years of Nonlinear History* (DeLanda, 1997: 58–60), opposites are not simply brought together, but, on the contrary, participate in the same process at a diagrammatic level.[17] On this pre-individual, isomorphic plane, bodies are replaced by 'organ-less', im/material relations whereby substances and forms lose their solid distinctions and flow into each other. In this indiscernible zone, a virtual idea emerges that transcends any specific or actual archipelago, whilst being inherent in all of them, at the same time.

To attend the theatre as an 'archipelagist' is to think of seas, of boats, of moving between islands. It is to be a body buffeted by winds, made seasick. Whether one is conscious of it or not, it is to inherit an Artaudian view of performance. For in the theatre, experience is always doubled, caught between landing and departing, an exchange that takes you elsewhere, a cruelty that ungrounds. In the drift of performance, bodies are actual and virtual, land and sea, mineral and wave. They stream and stand still. The stage is a *plateau* that moves in the manner of a tectonic plate.

In this drive to ecologise theatre, to make it archipelagic, Lyotard, as ever, is an insightful guide. In the short book *Enthusiasm: The Kantian Critique of History* (2009b), he says that the archipelago stands as a symbol for the faculty of judgement, which, though having no object itself, is able to move between the other faculties – cognition, ethics, desire, taste – in such a way that connections between them are made:

> [T]he faculty of judgment would be, at least in part, like an outfitter or admiral who launches expeditions from one island to another sent out to present to the one what they have found in the other, and which might serve to the first as an 'as-if' intuition to validate it. This force of intervention […] has no object; it has no island but it requires a milieu, namely the sea. (Lyotard, 2009b: 12)

To make a judgement, Lyotard reminds us, is never to stand apart from an object, to make it correspond to a pre-existing set of discourses or laws; rather, it is to plunge into an elemental force-field and to produce forms and worlds (actuals) through the very act of relating itself, of being willing to launch new expeditions, to take to the surf. Glissant concurs:

> Let us repeat this, chaotically: Relation neither relays nor links referents that can be assimilated or allied only in their principle, for the simple reason that it always differentiates among them concretely and diverts them from the totalitarian – because its work always changes all the elements composing it and, consequently, the resulting relationship, which then changes them all over again. (Glissant, 1997: 172)

As shape, figure and mode of expression, the archipelago does not protect you from the chaos or cruelty of the sea, from the virtual touch of its spume. It implicates you in the pre-individual flux, telling you that there is no escaping the ever-changeable systems of wind and water that makes every journey singular and infinite. Yet how exactly does one navigate the treacherous seascape of the stage? What trajectories to take? Where to place one's gaze? For in keeping with Glissant's and Lyotard's suspicions of the concept, not every archipelago is ecological, some are gulags, land-locked islands, forged in a logic of insularity, looking to maintain territory – the very things that theatre ecology wants to escape.

In response to the dangers above, I offer a set of conceptual provocations – 'Archipelagic fragments' – for making progressive ecological journeys. The fragments that I present, with headings in italics and centred on the page, are reworked expressions, elliptical instances of the ecological idea that I have tacked and drifted with throughout this book. In the attempt to outwit the tyranny of the concept, its totalitarian dryness, they are intended as acts of archipelagic communication in and by themselves, islands that express the kinds of vertigo which theatre seems so adept at disclosing. The writing is both provocation and evocation. From it, one should get a sense of a theatre ecology that is no longer confined to the stage but at work in the minds and bodies of spectators, too. Movement is all, the primary thing. The aim is to encourage all who read these fragments to set sail, to become archipelagic spectators, willing to experiment with what the Artaudian scholar Joeri Visser terms 'crescive writing'. This is writing whose transmission of speechlessness evokes 'static', allowing for frequencies in which the clinical, creative, and political aspects of theatre ecology are experienced together and yet kept apart (Visser, 2021: 103).[18]

Archipelagic fragments

Gestus

The *gestus* is a prodding, a reaching out, a projectile. It gesticulates and retracts; it allows a gap to appear, a void that is not a void, an indiscernible passage. The *gestus* is in the gesture, but it is not reducible to it. Where the gesture repeats the familiar, the *gestus* undoes. It leaves us dumbfounded; sometimes it shipwrecks us. Always, it 'constitutes the impenetrability of the work to thought' (Lyotard, 2009a: 41).

To feel the *gestus* of theatre is to be attuned to bodies, to a somatic restlessness, an enervation, that resists being put into signs, an energy that wanders through those strange gestures that create archipelagoes. Confronted with this force, this im/materiality, the analyst no longer dissects the stage; they find themselves tossed beyond its shore, lost, a body amidst a sea of other bodies – in a streaming world, a cosmic theatre.

Cruelty

To become archipelagic is to live with cruelty, to realise that one can never provide a full account of one's experience or catch up with the work. Theatre is matter that moves, it never keeps its appointments. Its necessity is implacable. The passing of theatre takes our breath away; it disturbs; it suspends; it violates; it incompletes. It puts everything in relation. It temporalises: 'There is a latent cruelty in the sensible. The gesture passes over a work like a breath, it passes into it and there exhausts itself, it passes over to the other' (Lyotard, 2009a: 49).

Enigmas

Unlike Lyotard who locates the enigma in painting, Mario Perniola links it directly to theatre and its effects:

> Their effect is not stagily theatrical, but 'theatric', since they do not attempt a mimetic representation of the action, but focus attention instead on the particular physicality of the actor, regarding his/her body as a clothing of flesh, as a tunic of skin of some entity to which no adequate form could possibly be given. [...] There is no heart to such things. They are only surfaces and nothing else. (Perniola, 1995: 45–6)

Building on this, Lisa Robertson, writing about the sculptor Eve Hesse, links it to refusal and liberation:

> It is the refusal to be defined (as, for example, a woman, as a painter, as a German), which is the basic liberatory gesture; this refusal opens a fantastic negative space – the non-yet, which rests beside and other than the question of an identity designation, without entirely eclipsing it. (Robertson, 2012: 43)

Glissant, by contrast, finds it in opacity:

> Agree not merely to the right to difference but, carrying this further, agree also to the right of opacity that is not enclosure within an impenetrable autarchy but subsistence within an irreducible singularity. Opacities can coexist and converge, weaving fabrics. To understand these truly one must focus on the texture of the weave and not on the nature of its components. (Glissant, 1997: 190)

Part of theatre's power, its ecologic, resides somewhere in the archipelagic interplay between Perniola, Robertson and Glissant. In this 'imperceptible inching towards something that is different', the earth is present as a force of circulation, an energy flux, a texture that undulates (Perniola, 1995: 12).

Bodies

To speak of bodies does not just mean the actual bodies of actants – be they human or non-human. It encompasses im/material bodies, the flux of the pre-individual moving through space, circulating like a weather system, creating a sense of thickness in the air, charged ions, a storm at sea, a soft ripple. In that fleshy im/materiality, the universe is 'present' as a theatre and the theatre something to navigate and drift through. Theatricalised bodies '*are material entanglements enfolded and threaded through the spacetimemattering of the universe*' (Barad, 2014: 182; original italics). They diffract.

Refrains

Refrains are nothing in themselves. They are pure consistency, bands of force, sonorous shapings, blocks of sensation: repeats, rituals, rhythms. Refrains compose open worlds; they 'trip' systems and 'flip' routines. They make the world dance, a discontinuous continuity, a swarm of flies in the air, a thunderous mist of stuff. Refrains blend fragments as fragments. They merge; they orientate; they attune. We are in them and mobilised by them; they take us beyond the known. 'A song [that] sings the earth and signs a body' (Grosz, 2008: 50), an orphaned song, a cosmic song, the voice of the *oikos*.

Transference

More than mere vehicles for the communication of ideas, theatrical productions draw attention to the medium of communication itself. They make it shimmer. In this becoming 'expressive' of materials, spectators get mixed up in what they perceive. They lose their bearings. Incomplete relations are established, compelling new embarkations. To sail through an archipelago is to find oneself face to face with islands that sometimes swell up unexpectedly like shapes in the mist.

Infection

To tack the process of transference, the performance analyst must be affected by brine, the smell of the sea, the pull of strange shores. Theatre folds into and draws from the non-human; the theatrical desire that matter has to be always other than what it is, to commingle and compose, to contaminate itself.[19] The transference of theatre ecology is not a release from; it is a rupturing into; a becoming mutant, a disparity, something larval – an unselfing.

Transversal

The spectator is invited to trace the staggerings of a transversal line from the sensations of theatre things to historical and political events, using the skin as map and thread. But always with the caveat that the intelligence should come after the enigma. It unfolds from the resonance of a mass of moving gestures, clashing and colliding in the surf of performance. The transversal is creative and political at exactly the same time: 'We will only escape from the major crises of our era through the articulation of a nascent subjectivity, a constantly mutating socius, an environment in the process of being reinvented' (Guattari, 2000: 45).

Passibility

Passibility is a medieval word. It merges possibility and passivity, allowing things to pass through with a gentle force. Lyotard uses it to describe an aptitude for being affected physically by the gesture of the work, for being indisposed to it, remaining open to the de-phasings of its *gestus*. The passible makes a theatre ecology possible because it teaches the spectator to withstand what sailors called the doldrums, those windless latitudes and longitudes where nothing can be done, where the pre-individual circulates in the lingering materiality of things.[20]

Experiment

In most Latinate languages, the verb 'experiment' means both to feel and to try out, to test something in the most empirical of ways. An experiment is always corporeal, an affair of matter's rippling outwards; its ecstasy. To practise theatre ecology is to experiment with boat building, learning how to elaborate, to adventure with a work.

Showing up

In a very beautiful essay on Beckett's *Happy Days* (1961), Joe Kelleher proposes that Winnie's ecology is found in her willingness to show up, each day and each night, to put herself to the test of theatricality, to sit under a burning sky (Kelleher, 2015b: 143). She is so passive she becomes active. In her attendance, she transforms her heap of sand into a desert island, a place, then, to begin again, to imagine the sea, to be out-of-step with one's self – an ec(h)ology. There is no hope here. All that rigmarole has gone. What is left is the 'happiness' that comes from surrendering, a becoming mineral, sinking into her dry garden.

Anamnesis

Unlike voluntary memory, which is historical and volitional, anamnesis is the memory of something that, while it has never been present, has also never left us. It confronts us with the 'immemorial time-space of the *oikos*'. But what does the anamnesis of theatre consist of? What impossibility do we re-member through it? Maybe this: an intuition of matter's virtuality, the sense in which as spectators, as humans, we owe fidelity to this invisible force and flow, this newness that is never new, at least not quite. Theatre's anamnesis happens when it discloses itself as an im/material flow, a cosmos in movement, a sea with stars.

Diffraction

To be diffractive in one's analysis is to meet the performance half-way; to be 'ecologised', an experience that is set in motion when the analyst submits to the shadowy movement of the *oikos*, when they relate and relay through an archipelagic sea, when they open themselves to the infinite, to everything that escapes interpretation.

Writing

Exposed to the pre-personal, shadowy presence of the *oikos*, language too diffracts. It has to be alive to the weight and density of theatre's *gestus*; to become the thing that it cannot capture, to sabotage itself. Its 'resonance is endless. It is everywhere but it is not just anything' (Wood, 2014: 4). The shape of the argument, too, has to change. The imperative is both to look for islands, and to bolt for open water; to be at least two.

Practice

Theatre ecologists eschew objectivity. They realise that analysis cannot be applied in a transcendent or totalising sense. Any 'cuts' made, any interpretations proffered, will be indeterminate, equivocal, a becoming with and through the work at a particular space and time, a diffractive intervention – something singular, something sympathetic, a way of differing, a new type of research. In the archipelago of the stage where everything is to be re-created; nothing remains the same. Only practices of ecology, openings to the *oikos*, modes of elaboration that are forever incomplete.

Constants

In this practice of assembling and reassembling three things remain constant:

- a willingness to be affected by the *gestus* of the work, to be implicated by theatre things: touch;
- a desire to elaborate ecological meanings, to attend to the incipience of theatre events: immanence;
- an obligation to fashion an expressive discourse, to express the inexpressible: possibility.

Scores for becoming archipelagic

- *Learn how to navigate*
- *Go beyond the skin*
- *Breathe with the eye*
- *Look with the lung*
- *Attend to the virtual*
- *Trip the refrain*
- *Affirm transference*
- *Surrender to cruelty*
- *Construct relays*
- *Choose not to choose*
- *Live the excess*
- *Affirm the surplus*
- *Wager on geometries*
- *Implicate and elaborate*

Wake

At a time when the planet is convulsing politically, demographically, environmentally, many of us in the Global North have forgotten the sea, lost touch with the sadness of its 'wake' – that still visible trace of a wave, which, for Christina Sharpe, as for Glissant, is inseparable from 'the continuous and changing present of slavery's as yet unresolved unfoldings' (Sharpe, 2016: 22). Artaud in his writing on pre-Colombian Mexico calls the 'wake' the Judgement of God; Anna Lowenhaupt Tsing (2015: 36) and Donna Haraway (2016: 99–100), closer to Sharpe, call it the 'Plantationocene', an ugly word for an ugly operation, forged in a logistics of transportation, standardisation and incarceration.

In the face of the Plantationocene, we, the landlocked, need a different approach to the 'wake' and to the *oikos*; one that would be archipelagic, that would encourage nothing less than a new ecology, a new politics and a new clinic. We could do worse than to listen to Achille Mbembe, who, after Glissant, calls for a pharmacy of the passer-by, a clinic of the one who drifts, who makes landfall where they will, where they can. 'Becoming human-in-the-world', Mbembe proposes, 'is a question neither of birth nor of origin or race. It is a matter of journeying, of movement, and of transfiguration' (Mbembe, 2019: 187).

In its own modest and minor way, theatre can help create this 'pharmacy of the passer-by' by transforming itself into a boat for journeying through an earth that it always contains within itself as a pre-individual force – a sea – that takes identities to the brink, to the point where people and things become indiscernible. This is a theatre ecology that moves, remembers and becomes, that stays in and with the 'wake'. But prudence, as ever, is needed. Not all bodies are the same, and some journeys are more arduous and dangerous than others. Attending is key; patience, too. No one knows where the storm of the future will fall, nor what it may yield. It can wreck our boats.

'A smooth space will never suffice to save us', Deleuze and Guattari warn us (Deleuze and Guattari, 1987: 500), and neither will the archipelago. So many archipelagoes, Glissant says, have been 'laden with palpable death' (Glissant, 1997: i). And yet even here, in the haunted present of those deaths, the imperative is to set sail, again, on the 'thousand channels' that Glissant tells us lead to 'the one way ashore' (Glissant, 1997: 1). While I wager on the theatricality of Artaudian theatres, archipelagic thinking ultimately encompasses all theatres; the potential is there any time a body appears on stage. A desert can be as archipelagic as a sea, and so, too, can a theatre, a *plateau*. What really matters is to construct a new mode of relating. One that is founded on the vertiginous realisation that the earth itself is archipelagic, an elemental flux of opaque energies, plate tectonics and technologies, drifting in 'undecipherable magma' (Glissant, 1997: 164). To respect the enigmatic performance of that submarine magma, to feel its intermittent pulse and flow in the human realm, on the space of the stage, is what my idea for a theatre ecology is so concerned to tune into and relay.

In the final pages of this chapter, I give readers a taste of what that ecology consists of by providing two short, archipelagic accounts of performances by Kris Verdonck and Gisèle Vienne. These accounts are not only summations that look back; they are intended to point forward to the modes of analysis and type of writing that I adopt in the two volumes of theatre ecology that follow this one.

Kris Verdonck's *End*, Kaaitheatre, Brussels, 2008

In *End*, the stage is divided into two levels: a raised background set about one and a half metres above the floor and a flattened foreground covered with what looks like mud or ashes – it is impossible to tell which.

The air is full of falling debris.

The performance works according to a logic of accumulation in which a staggered line of lonely bodies exit and enter randomly but persistently on both levels, each following its own stubborn path or track across the stage. Nobody communicates. Dialogue has been abandoned. A factory of alienation. Industrial theatre;

like a conveyor belt, the movement in this hellish landscape is always left to right;

the movement is relentless;

the performance starts with a man in a suit dragging a heavy load – an engine – that drives the whole apparatus. The man stumbles yet refuses to give up. We hear him breathe, a work horse. As he struggles across the stage, a bank of clouds moves behind him on a screen. A kind of joke, a bitter one, metatheatrical. Like theatre, clouds never stop: it is impossible to fix them; they live to mutate;

then another figure emerges – a man with white hair (Johan Leysen) – seated in a Perspex recording booth, speaking a litany of disaster from an assemblage of found texts, intoning them into a microphone as the booth makes its slow way across the stage;[21]

Achingly;

then another enters;

but this time from the sky, thudding into the mats on the raised platform, before getting up and exiting, holding a cotton bag in his hand and moving this time right to left, in contra flow;

above them, another body now – a kind of bird man – suspended from the ceiling, in a harness, helpless, clumsy, a creature out of its element;

then two women enter;

one is well-dressed, in an office worker's suit, glittery heels and blonde wig. Her movements defy gravity, a marionette attached to visible strings, manipulated by some unseen hand.

She is pregnant;

the other wears an elegant blue dress and fur coat. She is in a frenzy, running, always looking back, lugging a body bag that she refuses to relinquish;[22]

there are non-human bodies, too.

A line of fire – burning phosphorus – blazes white producing chemtrails and clouds of smoke; a loudspeaker – a choir – blasts out military music and human wailing, a soundtrack for the chaos unfolding in front of us; and then, towards the end, a petrol engine hanging from the rafters, buzzing aggressively, a monster that stinks and reeks;

judgement day. A constant painful, drone in the ears;

the stage is dimly lit, shadow mostly;

along with the fire, and the reading lamp, light spills from the projections of clouds, illuminating falling matter, airborne particulates. This is performance as cataclysm and ruin, a kind of aftermath, the moment when history dissolves and fragments are all that remain;

western modernity: the time of the machine; a toxic temporality;

the man in the transparent booth – the only one with words, borrowed ones – says:

> Nobody really knows what is happening. Nobody [...] The disaster has lodged itself in our bodies and genes. It is propagating itself. It is our legacy. (Verdonck in van Kerkhoven and Nuyens, 2012: 247, 250)

At some point in the performance, two birds are released;

nothing is said about ecology or 'Nature' in this performance. Everything simply evokes the coming end of the planet as a transversal, a line that joins disparities. On that line, through that geometry, politics, aesthetics, and environment shoot through each other. It's impossible to separate anything out;

history is everywhere. Time and space overlap. They spill over; they fissure; all co-ordinates collapsed. But to be left with what, exactly? Only this, a virtual world that haunts the so-called real one, the one predicated on Newtonian physics, the one that is finishing;

something else, something other, is revealed by these bodies in distressed transit. Something we didn't want to hear; something we didn't want to feel.

It's as if the chaos of the Anthropocene was given shape, launched across the stage, impacting on us, showing no mercy. Leaving only suffering. And yet in that suffering, in the violence of its rhythms, the cruelty of its drones and textures, *End* performs the ecology that other more committed works are concerned merely to speak about.

This is ecology as *oikos*, ecology as shattering, ecology as negation. The bodies have expressed it all. Spookily. Virtually. Transversally. In their hum and resonance. By vibrating together apart.

Nothing was ever named in this loud cry for a new people and a new earth. And yet everything was felt.[23]

Gisèle Vienne's *Crowd*, Tramway, Glasgow, 2019

In a theatre – on an earth – covered in mud, a ragged column of performers enters the stage. Sometimes the performers enter alone, sometimes they come in two and threes. Music – a techno-beat, a dream sound – washes the auditorium.

The music – this recorded music – creates a skin, a mobile surface, a membrane that pulses through us, through all us, uniting us in disjunction, an impossible plenitude that moves, a million heartbeats synched in asymmetry, or so it seems.

One can never be sure....

The performers walk in slow motion; they move like GIFs, broken up in their flow, staggered and disturbed. In this dance, this syncopated avalanche, they are all gesture, folded in on themselves and yet entangled with all the others. Digital blood.

They reach out to us and yet they retreat. This is why we follow them. This is how we are with them. I no longer know where I am. Where do the bodies stop? What are my limits? Why am I moving?

In this place of collapse, meaning drops away, and in that absenting of sense, we drift into a gap, a kind of sea.

Something comes towards us; something retreats. Nothing is deferred in this oscillation and yet no object emerges.

Something is in the process of being born, an indiscernible something that reorientates by disorientating, that seems to emerge from the rhythmic gravity of bodies that are simply there, plugged into some invisible ecology; one that ripples through time and space as a current that cannot be seen, a great wave of becoming, organic and inorganic, material and immaterial.

Through this molecular wave, in this mobile interstice, a new way of being discloses itself in everything that language cannot say. It comes all at once. Like the earth, this ripple 'touches' but escapes any grip (Lyotard, 2009a: 121).[24] It is always just passing through; on its way to some elsewhere, a secret.

Notes

1 I explain the difference between a method and a practice on pp. 190–203 in this chapter. But for now, I will simply say that whereas a method is generic and stable, a practice is specific and mutational. It changes depending on the context.
2 In *To Watch Theatre: Essays on Genre and Corporeality*, Fensham does not make good on the environmental resonance of her maritime metaphor. In that text, her concerns are reserved for the human realm alone. 'To watch theatre',

she insists, is 'to watch carefully for the remaining signs of a fragile humanity' (Fensham, 2009: 23). In recent years, however, she has started to look beyond 'fragile humanity' and to engage with the fragility of the earth. I am thinking, particularly, of her editorial contribution to the 'On Climates' issue of *Performance Research* (2018) and her essay 'Ecological Combustion: The Atmospherics of the Bushfire as Choreography' (Fensham, 2020).

3 Glissant states that 'in Relation every subject is an object and every object a subject' (Glissant, 1997: xx) – by which he means that every description is part of the thing it describes, mixed-up in it.

4 In the same way that there are different types of theatre, so there are different modes of performance analysis. One would not, for instance, expect to do a semiotic analysis, necessarily, on certain forms of performance art or sculptural modes of site-specific theatre. For more on this, see Patrice Pavis's conclusion to *Analysing Performance, Theatre, Dance, and Film*, entitled 'Which Theories for which Mise-en-Scene' (Pavis, 2003: 270–308).

5 According to Brecht: '"Gest" is not supposed to mean gesticulation: it is not a matter of explanation or emphatic movements of the hands, but of overall attitudes. A language is gestic when it is grounded in a gest and conveys particular attitudes adopted by the speaker to other men' (Brecht, 1964: 104). Where Brecht is always interested in the 'social gest', Lyotard is concerned with the capacity of *gestus* to escape meaning and to open subjects to an enigmatic reality that the social is unable to fathom (Brecht, 1964: 104–5).

6 In the text 'Gesture and Commentary', Lyotard states that the *gestus* of the work of art is defined, above all else, by this 'absolute insignificance', the refusal of the work to give up its secret (Lyotard, 1993b: 38).

7 See, for instance, the essays 'Scapeland' and 'Domus and Megalopolis' in the collection *The Inhuman: Reflections on Time* (Lyotard, 1991), both of which exist as companion pieces, in many ways, to '*Oikos*'.

8 Anamnesis is neither memory nor forgetting (amnesia). It is the return of a feeling or affect that has never entered consciousness as an image or object.

9 In *The Inhuman: Reflections on Time*, Lyotard notes that we should 'love' matter, and that certain art forms – he cites those of Duchamp and Mallarmé, as exemplars – allow for its anamnesis, the dizzying rediscovery of a creative earth (Lyotard, 1991: 46).

10 Nilling, for Robertson, is close to the word nihilism. But with the caveat that the *nihil* is understood as a gap, a type of writing that punctures language to open a space – a void – for something else to emerge.

11 Advanced by Tim Etchells (1999: 17–18), who, perhaps, unconsciously draws from Jerzy Grotowski, the figure of the witness was much cited in Theatre and Performance Studies in the 2000s by theorists and practitioners alike. In the main, the term was used to designate a type of ethical spectator whose comments on, or writings about, performance were needed as 'proof' of the fact that the work had taken place. As both Patrice Pavis (2016: 25–6) and Rachel Fensham (2009: 14) have shown, witnessing does not cover the complexity involved in what it means to 'watch theatre', in terms of affect, co-creation and politics.

They are also dubious about its relationship with catharsis. Lyotard states that 'in witnessing, one also exterminates. The witness is a traitor' (Lyotard, 1991: 204).

12 For more on 'working through', see the section on transference in Chapter 3, pp. 114–15, in this book.

13 As Franco Berardi explains in his parsing of Guattarian transversality, conjunction is an exchange based on sympathy, in which bodies are allowed to drift, to become other, to take unexpected journeys, as Lisa Robertson did when looking at paintings in her grandmother's house (see pp. 187–8, in this chapter). Connection, by contrast, is an action that 'concatenate[s] bodies that have been codified or formatted' in advance and that always stay the same (Berardi, 2015: 21). Where conjunction is excessive and transformative, connection is synthetic and restorative – it changes nothing. In it, subjectivity loses its existential freedom to break with those systems that oppress it.

14 Kathleen Stewart's work on affects and atmospheres has also been a major influence (Stewart, 2007).

15 There is a close connection between Glissant's archipelagic thinking and Walter Benjamin's notion of the constellation. Both seek to bring things together in a non-totalising manner.

16 To see the earth as a globe is not only to close it off from its cosmic outside, but it suggests that it can be viewed holistically, consumed as a stable image. The result is sameness, oneness, totality. Globalisation is also problematic when considered in its economic sense, as a process that subjects every locality to the supposedly universal law of capital exchange value.

17 My position on metaphor is close to that of Isabelle Stengers who in 'Reclaiming Animism' sees it as performing a 'safety function', protecting us from dangerous contaminations, from what Artaud would call the plague (Stengers, 2012).

18 Visser uses the term to apply to Artaud's writings. 'Crescive writing', as he has it, 'is critical creative writing or better, an affective and cruel writing' (Visser, 2021: 103).

19 Claire Colebrook tells us why: 'Once sensation is liberated from a produced relation or sensation as it is lived, we reach [through art] sensation as it stands alone, the vibration of power to differ from life, from which relations are effected: sensation in itself' (Colebrook, 2006: 99; my addition).

20 See also how Jean-Luc Nancy uses the term in *The Gravity of Thought* to describe 'a passivity [...] that cannot be determined in opposition to activity' (Nancy, 1997: 69).

21 The texts are a collage of writings from Lord Byron, Alexander Kluge, Curzio Malaparte, Richard Rhodes, W. G. Sebald and internet reportage.

22 In the text *Listen to the Bloody Machine: Creating Kris Verdonck's End*, Verdonck explains how the figures were given names and associated with historical events: the man in the booth was the messenger; the man who drags the machine, Stakhanov, the heroic Soviet worker; and the falling man, the Ludd. The woman with the body bag was associated with the city of Sarajevo, and the birdman with those who jumped from the top of the World Trade Center on

11 September 2001. For more on their stories, see van Kerkhoven and Nuyens (2012: 57–85 and 141–90).
23 For more on the ecology of Verdonck's theatre, see Lavery (2020).
24 Erin Manning deals extensively with a similar idea of touch in *Politics of Touch: Sense, Movement, Sovereignty* (2006).

Afterlude:
double cut/frozen wave

Differently from many existing publications in the field of theatre, ecology and the environment, this book was conceived as a theoretical project, an attempt to advance a new idea of theatre ecology by reconsidering the ecological affordance of the theatrical medium itself. Rereading it now for the purposes of what can only ever be a tentative conclusion, I realise that there is a manifesto-like quality to it, a desire to critique extant ways of thinking about how the environment has been conventionally written about and analysed. In the polemical, early chapters, the imperative was to encourage a departure from dominant methods and theories of doing ecocriticism within Theatre and Performance Studies and Environmental Humanities; in subsequent chapters, the intention was to shift the focus from interpretation to action, from reading to writing within the milieu of the theatrical event.[1] In both instances, the goal was to create a new vocabulary for considering western theatre as a terrestrial art form, an ecological event that one is unable to stand apart from and dissect at a distance. In making these ecocritical moves, in affirming the immanent ecology of theatre that was there from the early Platonic critiques of the medium, I have looked to show how ecology is a politics, aesthetics and ontology that reconfigures the limits of subjectivity itself. For how can one ever hope to redistribute ecological and planetary agency without questioning the desire, the will-to-power of the subject doing the redistributing? Not to do so is to run the risk of perpetuating a mode of existence, an ecological *habitus* that was fundamentally flawed – toxic – from the very beginning.

Conceptually, one of the key lessons that revealed itself to me through the process of writing this book, an epiphany I would like to insist on, is that everything *really* does happen on the margins, in the interstices, in how relations and relays work together. In proximity to these 'contact zones', it is not enough merely to reference tropes such as borders, thresholds, entanglements, mixity, liminality and intra-activity. Rather one is always obliged to say what those median words mean, to appraise how they are constituted, to contemplate their operations and affordances, to find ways of becoming with and from them. To recall Claire Colebrook's caution one last time, some forms of posthumanism unconsciously (and tragically) perpetuate humanism's expansive dream of oneness, the overcoming of what Simondon calls the 'ambivalent form of tension and incompatibility' of the living being (Simondon, 2009: 15). Faced with these inadvertent but dangerous repetitions, it is incumbent on the ecocritic to be as accurate as possible in one's thinking of posthuman relationality, to try to point out blind-spots and possible danger points in advance, while all the time realising that not every base can be covered. Something always escapes the grip of intentionality, leaks out. Such is cruelty, such is virtuality, and such, too, is the ambivalence of the *pharmakon*. But to renege on this drive for criticality is a serious matter. At best, it perpetuates a useless idealism, an empty performative; at worse, it results in the cynicism of the professional critic and practitioner. In both instances, things deteriorate and worsen, the negative aggregates.

Confronted with the magnitude of the ecological, economic and political pressures currently weighing on the planet, something else is needed in the Environmental Humanities and Theatre Studies; something that would challenge the sovereignty of the *anthropos* from the inside, that would be ontological as well as political – something 'radical', then, if that exhausted, overwrought word still retains currency. In these pages, the word radical has been used in its proper etymological sense to convey an action of uprooting, which, in my case, as in Erin Manning's, is targeted at 'the volition-intentionality-agency triad' that subtends dominant western humanist ways of conceiving what the individual self supposedly is and how it operates (Manning, 2016: 6). Gilles Deleuze and, especially, Félix Guattari acted as important guides for this project when they evoked the need to construct a different earth and a new people to inhabit it, a process of re-singularisation.

In essence, all I have been trying to achieve in this theoretical book is to describe a journey towards this singularity, this people, this earth; to track a process in which theatre becomes an ecological nerve machine, whose irritations disclose an existential wound within western subjectivity – what Lyotard terms the *oikos* and Artaud names cruelty. This is done not for the purposes of nihilism or to gratify some misguided, atavistic urge, but in the name of an ecology whereby aesthetics, politics and subjectivity would exist as a transversal nexus, a way of being 'cut together apart'. The stages of that journey, along with construction of that nexus, form the two axes of what I have been calling 'an idea of a theatre ecology', the somewhat austere and perhaps old-fashioned title of this book. The great irony in that title – at least for readers who were expecting the explication of some Platonic or Kantian notion of conceptual thought – is that my idea has nothing intelligent or rational about it; no desire to be wise and autonomous. On the contrary, I have looked to use the idea against itself, to argue for the potential of feelings and forces that exist beyond the logical mind, that come from the non-human movements of an always already theatrical ecology – a bastard, creative flux that I have referred to throughout this book as the virtual, the incorporeal, the im/material, the pre-individual, the *oikos*. Theatre ecology is an impossible idea. More of a dream, really: a desire to reach the metastable space before thinking, 'that by which the given is given' (Deleuze and Guattari, 1994: 139).

Ultimately, it is not enough to explain what theatre ecology is, no matter my attempt to do so in this volume. The essential move is to gesture towards what it might become, to express its potential, to exemplify it. In that respect, and as I impose a 'double cut' that will simultaneously sever and suture this text from the two volumes to come, it is perhaps most productive to figure my 'idea' as a map of sorts. Not a 'road map' that would scope out the ground, skirting the territory in order to tame the unknown, but rather a tentative, virtual map, a type of 'transductive' sketch, giving birth to itself in the very act of trying to navigate a terrain. Some will disagree with it, others will reject it, but hopefully there will be those who respond to it, taking it in different directions, tacking it in their own ways, with their own contexts, commitments and lexicons in mind. In the same way that the 'idea' is always out of step with itself, so the fate of this book is out of step with my own desire to pin it down. And that is exactly as it should be.

Ecology has been accorded multiple meanings in this book. It has been described as an event, a process, a becoming, an affect, a virtuality, a terrestrial force. At all times, though, it refers to something that cannot be measured or controlled, a type of theatre that simultaneously worlds and un-worlds. To turn to Simondon, ecology 'phase-shifts', it shows life to be unstable and errant (Simondon, 2009: 6). As a 'theatre of individuation', ecology is a surplus that disturbs and disquiets the identities of the bodies that host it – hence the centrality of Artaud to this project. For of all twentieth-century practitioners in the West no one did more than Artaud to highlight the necessity of ecology, to posit a terrestrial stage. Where the Dadaists and Surrealists before him talked of viruses and microbes, of the need for theatre to materialise itself, Artaud was the first to tie what he called 'the plague' to the earth, to argue for an ecology that would contest the biopower of the US military-industrial-agriculture machine. To get to Artaud's ecology, it is necessary to emancipate him from a number of extant readings that in their legitimate drive to critique the neo-colonialist thinking and language of his project have nevertheless tended to underestimate its relevance, to sideline its ecological potential in the here and now. As I showed in Chapter 5, it is only by placing Artaud within a history of capitalist despoliation that it becomes possible to get some sense of his contemporary resonance, of the ecological problematic he has left us with: namely, how to get beyond the deathly spell of capitalist sorcery, its malfeasant desire to exit the earth?

Artaud's pharmacology is predicated on the assumption that there is no absolute divide separating 'Nature' from 'Culture', that theatre is ecological because ecology is theatrical. In Chapters 2 and 3, I sought to unfold the stakes of that doubled ecology by introducing a theoretical vocabulary, based on the writings of Édouard Glissant, Gilbert Simondon, Karen Barad, Jean-François Lyotard and Félix Guattari. As the book advanced, other thinkers, such as Michel Corvin, Josette Féral and Samuel Weber became part of that assemblage, too. My intention in bringing them into Artaud's orbit was never to show how his theatre conformed to their ideas, but rather to repurpose their thinking to illuminate the ecological affordances of theatricality itself. That allowed one to see, in other words, how theatre could be conceived in Artaudian terms as an ecological apparatus, and not as an instrument for narrative, theme and reflection.

The space needed to develop that proposition has meant that I have not dealt in detail with specific productions, apart from two short accounts of work by Kris Verdonk and Gisèle Vienne. In similar vein, there has been no specific engagement with classes, genders, colours or sexes. More will be revealed on these fronts in volumes 2 and 3. Such absences were not made in the name of some reductive universalism, an unmarked and assumed human essence where all differences would be reconciled and overcome in phallogocentric whiteness. Rather, I was holding out for a different kind of abstraction, the necessity of gesturing towards the 'organ-less', pre-individual plane where a new logic of affiliation can be encountered and a possible future born. An indiscernibility that would provoke the slippage needed for a new becoming, a way out.

That this indiscernibility, this pre-individual flux, is necessary for eco-political transformation is not only evident in the importance I attached to Guattari's transversalist rethinking of transference, the intensive break that, for him, deterritorialises the overcodings of capitalist semiosis, allowing desire to flow again. It is also found in my references to the postcolonial ecologies of thinkers such as Édouard Glissant, Malcom Ferdinand, Achille Mbembe, Christina Sharpe and Paul Carter who are for a new planetary politics, founded on an alternative metaphysic to the 'white' one that Artaud associated with God. Consider this from Glissant and then relate it to what I have just argued for:

> The opaque is not the obscure [...]. It is that which cannot be reduced, which is the most perennial guarantee of participation and confluence [...]

Afterlude 211

[T]he thought of opacity saves me from unequivocal courses and irreversible choices. (Glissant, 1997: 191–2)

In *Theatre and Landscape: Taste, Ecology and Politics*, and *Theatre as Clinic: Constellations of the Anthropocene(s)*, I actualise the potential inherent in opacity by focusing on the work of a diverse body of practitioners all of whom, but in different ways, approach ecology as an aesthetic and politics, a mode of relating that troubles distinctions between human and non-human worlds. Some of these artists go by the names of Vinciane Despret, El Anatsui, Basel Abbas and Ruanne Abou-Rahme, Claudia Fontes, Ana Mendieta, Samuel Beckett, Ivana Müller, Kris Verdonck, Alistair McDowall, Philippe Quesne, Ibrahim Mahama, Faustin Linyekula, Sheila Ghelani and Sue Palmer, Kazuo Ohno and Min Tanaka. In the theatre and performance archipelagoes that these theatre ecologists create, spectators are invited to drift through the matter of performance like a ship drifting through a sea, in whose 'indefinite swell [...] all reality is dissolved' but where history still holds sway (Mallarmé, 2006: 179).

As my theory of archipelagic spectating in Chapter 6 demonstrates, the style of writing associated with this duplicitous riptide is not only opposed to the standard language of academia, it is expressive of the alternative activism of the artwork I discussed at the very beginning of this book. This is an activism that renounces policed distinctions between activity and passivity, heteronomy and autonomy, and virtuality and actuality. It is an activism, too, that cleaves tight to theatre's capacity to provoke bifurcations, to exist as the irritant or impurity that exposes critics and spectators to something – an *oikos* – that they can never grasp but only express. That there is no scientific way to measure the efficacy of this *oikos* is beside the point. Indeed, it *is* the point. The future is in the present, and the horizon is already here. Much better then, to remain faithful to theatre's ecological ability to effect a double cut, to set spectators adrift

in the swell that accompanies the wake on the cusp of this book's still
in the swell that accompanies the wake on the cusp of this book's still
in the swell that accompanies the wake on the cusp of this book's still
in the swell that accompanies the wake on the cusp of this book's still
in the swell that accompanies the wake on the cusp of this book's still
in the swell that accompanies the wake on the cusp of this book's still
in the swell that accompanies the wake on the cusp of this book's still
in the swell that accompanies the wake on the cusp of this book's still
frozen wave ....

Note

1 There is a shift between past and present tenses in this chapter. While some may find it ungainly, ungrammatical, even; I find it necessary. No conclusion is ever an ending. The present percolates through it, and all looking back takes place in the impossibility of the 'now'.

Glossary

Abhumanism: a concept taken from the theatre scholar Michel Corvin who uses it as a dramaturgical term for explaining the European avant-garde's rejection of Aristotelian representation in the mid- to late nineteenth century. Where Aristotelian theatre assumes a specific essence to the human that can be unproblematically represented, abhumanist theatre claims that the human is a plastic, malleable creature that is entangled with the non-human things (objects, props, languages, characters etc.) it shares the stage with.

Actual: that which 'is' – objects, forms and substance that have a fixed shape and 'individuated identity'.

Affect: non-human energies and sensations that are experienced by bodies. Affects are powers; they have the capacity to provoke becomings. They move everyone involved in the theatrical event.

Animism: a way of perceiving the world and the 'things' in it as agentic, lively, capable of creating affects and provoking actions at a distance.

Anthropocene: a term used by earth scientists to designate a new epoch in the earth's geological history, one in which history and natural history have lost their separateness and come together as a single dynamic force. In the Anthropocene, the human is conceived as a geological agent, a creature that has left its mark on the earth, but also a creature that is tied to the destiny of the planet. To exist as a geological agent is to realise that one is fated to die out once the earth has entered a new phrase in its geological history – hence, the ambivalence and irony in the so-called 'age of man', which is what the Greek etymology of Anthropocene stands for. The Anthropocene is often criticised as a term, since it assumes that there is only one humanity and that all humans are equally responsible for the transformation of the planet.

Anthropocentrism: the attempt by humans to see the world in their image by attributing human characteristics to diverse forms of non-human life (animals, rocks, plants etc.).

***Anthropos*:** another name for the human, but one that stresses its earthly destiny, its creatureliness.

Archipelagic: a metaphor inspired by the geomorphology of archipelagoes and used to describe modes of thinking and spectating that are able to move between and hold together heterogeneous elements and disparate materials.

Capitalocene: a term introduced by Jason W. Moore in an attempt to historicise the Anthropocene, to show that not every human being on the planet is responsible for changing the climate and geology of the earth in the same way. For Moore, capitalist modernity is the primary agent of planetary change, in the extent to which its processes and mindset have posited 'Nature' as something that can be used, transformed and sold with impunity.

Correlationism: a word used by speculative realist thinkers to critique the Kantian idea that reality corresponds with human thought and consciousness alone.

Cruelty: Antonin Artaud's word for describing the disruptive affect of theatre on the bodies of spectators. For Artaud, cruelty is always double – the suffering it imposes offers the possibility of wagering on a new life, one that troubles the human's belief in consciousness, volition and agency.

Diffraction: a term taken from Karen Barad to explain how a body is never one. As a consequence of the double-split experiment in quantum physics, diffraction shows how an electron is both particle and wave at the same time. Diffraction highlights the extent to which reality is multiple, ghostly and virtual.

Earth: a process not an object, a movement rather than a globe, a dynamic entity, a matter-energy system that impacts on and is impacted by life. For Deleuze and Guattari, the earth is the Ur-agent of deterritorialisation. It is always on the move.

Ecologising: the act of making something – in this instance, theatre – ecological. To ecologise theatre is to approach the medium as bound up with the arrangement and expression of non-human objects, materials and affects.

Glossary

Ecology: a term invented by Ernst Haeckel in the nineteenth century and used by biologists to describe the functioning of non-human ecosystems or *Umwelts*. In the Environmental Humanities ecology has been expanded to account for the interdependent and relational dimensions of all life forms, including human societies. In its expanded, and often contested, contemporary usage, ecology is a way of thinking, perceiving and performing. It is also an ethics, politics and aesthetics. In this book, it is considered as an event, a way of troubling human exceptionalism.

Gestus: a term taken from the work of Jean-François Lyotard. Radically different from Bertolt Brecht's more specifically theatrical understanding, the *gestus* does not present itself as a social 'attitude' to be read; rather, it is a sensate charge, a thrust of matter, that exceeds the aesthetic framework and touches the spectator in corporeal ways that they are unable to understand. The *gestus* does not ask to be understood; it simply demands an affirmation of earthly existence.

Humanism: a philosophy or credo in which the human is championed as an exceptional creature, endowed with 'natural rights' that can be accessed through the proper use of reason. Humanism proposes that the essence of the human is to be rational and self-present, a subject able to know itself and confident in its capacity to judge and administer universal 'truths'. Although humanism is commonly traced to the early modern thinking of Erasmus in the sixteenth century before finding its apogee in eighteenth-century European philosophy, in theatrical terms it is already present in Aristotelian notions of mimesis. In its dramatic form, humanism is a mode of representation based on the assumption that spectators will be able to recognise, identify and empathise with subjects *like* themselves, even if those subjects are presented as 'exceptional' individuals, as they are in Aristotle's theory of tragedy.

Im/material: the presence of the 'slash' in this word brings together two conventionally opposed terms: the material and immaterial. Im/materiality describes the force within matter, the reverberation or resonance of a power that while it is born from and manifests within material bodies and objects is not reducible to them.

Individuation: a term taken from the philosopher Gilbert Simondon to account for the unpredictable evolution and immanent becomings of life.

Intra-activity: a concept from the philosopher Karen Barad to describe relations *within* bodies, not just between them. Intra-activity assumes that there is no definitive or absolute separation between life forms and that symbiogenesis offers a more accurate insight into evolution than autopoesis. From an intra-active perspective, identities and things are heterogeneous from the very beginning, fissured from within, 'cut-together-apart'.

Non-human: though it points beyond, the non-human is also within the human, a corporeal substance and pre-individual force that the subject is unable to master, know or identify with. The non-human troubles the desire of the human to be whole and self-identical, which is why, in this book, it is preferred to synonyms such as the 'more-than-human' or 'other than-human', both which imply a border of sorts.

Oikos: a term taken from the philosopher Jean-François Lyotard to express the non-human 'guest' that lives within the human, and which always works to trouble the equanimity and balance of its host. Where the etymology of the word ecology borrows its prefix from *eco*, the Greek word for home, Lyotard's provocative phrase posits the *oikos* (and thus *oikology*) as symptom of uncanniness, of not-being grounded in a place, identity or language. The *oikos* is an instance of the non-human within the human, a thing that merits respect, attention and care. Critically, Lyotard also contends that the *oikos* is the 'faculty' that allows humans to be affected by forces inherent in terrestrial matter and therefore able to exist as creative beings, creatures with the capacity to become other than what they are.

Passibility: a type of responsiveness that escapes the conventional oppositions between activity and passivity in the spectator's engagement with the artwork.

Pharmakon: a Greek word that translates both as scapegoat and poisonous cure. Pharmacological knowledge is ambivalent knowledge. The *pharmakon* troubles all fantasies of transcendence, oneness and absolute knowledge. It shows that every decision is double-edged, implicated in its opposite. The other is always already the same. There are no binary solutions to existing on the earth.

Plantationocene: a term that builds on postcolonial and decolonial thought in order to contextualise and historicise the Anthropocene. The Plantationocene traces the root of anthropogenic climate change to the beginnings of European colonialism in the late fifteenth century and to the subsequent forced migration of Africans to the Americas and Caribbeans where they were sold into chattel slavery and set to work on colonial plantations. The Plantationocene also refers to settler violence and the genocide of First Nation peoples. The Plantionocene is rooted in colonialism, Empire and white geology. It describes an economic, political and environmental catastrophe. However, thinkers such as Glissant, Wynter, McKittrick and Yusoff have also been concerned to highlight modes of local resistance to the Plantation. In doing so, the Plantation is not a one-way street of colonial power; the refusal of enslaved subjects to undergo its violence harboured alternative ways of being on the earth and existing as a human being.

Posthuman: a word, like ecology, with a number of competing definitions. In this book, posthumanism is not a stage of development, a phase that comes after humanism, and neither is it associated with technology and computerisation. In keeping with Clare Colebrook's paradoxical definition, posthumanism predates humanism; it stands for an ontological rift within human subjectivity – a type of wound – that cannot be healed and which has been there from the beginning. Importantly, there is a politics to posthumanism, one that respects the capacity of human and non-human life forms to express themselves and to experiment with their potential to become other than self. To be posthuman is not to be inhuman. It is to undergo a trial and assent to a process, a word that in French (*procès*) has both a temporal and an ethical dimension to it.

Pre-individual: an immanent energy or force which, in the philosophy of Gilbert Simondon, catalyses becomings, allowing individuated life to de-phase and bifurcate in the name of something new. The pre-individual is a power, not an object.

Terrestrial: the power of the earth to impact on and transform bodies in unpredictable and immeasurable ways. Bodies are terrestrial because they are open to the forces of a cosmic planet – weather systems, earth tremors, moons, suns, elemental materials. To be terrestrial is to be born contaminated. We do not exist in a world of relations without being affected. There is neither immunity nor impunity for the terrestrial body.

Theatre ecology: a compound noun that refers to a double process. On the one hand, the theatrical quality of ecology, the sense in which life exists as a 'theatre of individuation'; and, on the other, the always already ecological dimension of the theatrical medium, apparent in the metamorphosis undergone by any object or body when placed on a stage. Although they are not the same, theatre and ecology share a common quality: the ability to reveal the virtual potential of life and the capacity to affect bodies in actual time and space, to play on their metastability, to make them volatile.

Theatricalisation: a mode of representing that highlights the transformative power of theatricality in both the theatre and beyond.

Theatricality: a self-conscious mode of signification that draws attention to itself in the very act of communicating and making sense. By showing appearance, theatricality undermines the 'truth' or actuality of what is being staged. In doing so, theatricality creates suspensive acts that lack the capacity of the performative to produce 'realities'. Yet theatricality's undoing of the actual is not to be equated with nihilism or nothingness. On the contrary, it allows for another reality to be perceived, one that is affective, intensive and virtual. For that reason, theatricality is best approached as a power or force. It embodies the ways in which life is always in excess of itself, a process of transformation and becoming.

Transversal: A geometric and clinical term used by the philosopher and psychotherapist Félix Guattari to describe a line that is able to assemble heterogeneous elements together without unifying them into a whole. As an instance of 'disjunctive conjunction', the transversal approaches ecology as a mode of being rather than as a delimited form of environmental action, a proper way of living with 'Nature'. Understood transversally, ecology is at once aesthetics, politics and ontology. Because the transversal brings things together in a dynamic nexus, a transformation in one domain can affect all others, which explains why Guattari places such faith in the power of aesthetics to provoke ecological change.

Virtual: a term taken from quantum physics to describe the expansive and heterogeneous potential that all forms of life possess. Virtuality is not opposed to reality; it is opposed to actuality, without, for all that, ever being separated from it. Every actual, in other words, is haunted by manifold possibilities. The virtual is a fullness, a plenum.

Bibliography

Adebayo, Mojisola, 2021 ongoing, *White Climate: African Literatures and Agri/cultural Practices*, Research Project, https://counterpoints.org.uk/event/agri-cultural-practices/ (accessed 20 June 2023)
Adorno, Theodor, 1997 [1970], *Aesthetic Theory*, trans. Robert Hullot-Kentor (London: Athlone Press)
Agamben, Giorgio, 2009, *What Is an Apparatus? And Other Essays*, trans. David Kishik and Stefan Pedatella (Stanford: Stanford University Press)
Ahmadi, Mohebat, 2017, *Towards an Ecocritical Theatre: Staging the Anthropo(s)cene* (PhD dissertation, University of Melbourne)
Ahmed, Sara, 2006, 'The Non-Performativity of Antiracism', *Meridians*, 7:1, 104–26
Alaimo, Stacy, 2008, 'Trans-corporeal Feminisms and the Ethical Space of Nature', in Stacy Alaimo and Susan Hekman (eds), *Material Feminisms* (Bloomington: Indiana University Press), pp. 237–64
Alaimo, Stacy, 2010, *Bodily Natures, Science, Environment and the Bodily Self* (Bloomington: Indiana University Press)
Alaimo, Stacy, 2014, 'Thinking as the Stuff of the World', *O-Zone: A Journal of Object-Oriented Studies*, 1, 13–21
Alaimo, Stacy, 2016, *Exposed: Environmental Politics and Pleasures in Posthuman Times* (Minneapolis: University of Minnesota Press)
Alston, Adam and Martin Welton (eds), 2017, *Theatre in the Dark: Shadow, Gloom, and Blackout in Contemporary Theatre* (London: Bloomsbury)
Alter, Jean, 1990, *A Sociosemiotic Understanding of Theater* (Philadelphia: University of Pennsylvania Press)
Andermatt Conley, Verena, 1996, *Ecopolitics: The Environment in Poststructuralist Thought* (London: Routledge)
Angelaki, Vicky, 2019, *Theatre & Environment* (Basingstoke: Palgrave Macmillan)
Arons, Wendy and Theresa J. May (eds), 2012, *Readings in Performance and Ecology* (Basingstoke: Palgrave Macmillan)
Aronson, Arnold, 1981, *The History and Theory of Environmental Scenography* (Ann Arbor: University of Michigan Press)
Artaud, Antonin, 1958, *The Theater and Its Double*, trans. Mary Caroline Richards (New York: Grove Press)
Artaud, Antonin, 1976, *Antonin Artaud: Selected Writing*, ed. Susan Sontag, trans. Helen Weaver (Berkeley: University of California Press)
Artaud, Antonin, 1992 [1947], *To Have Done with the Judgement of God*, trans. Clayton Eshleman and Norman Glass, in Douglas Kahn and Gregory Whitehead

(eds), *Wireless Imagination: Sound, Radio, and the Avant-Garde* (Cambridge, MA: MIT Press)

Aston, Elaine, 2006, '"A License to Kill": Caryl Churchill's Socialist-Feminist Ideas of Nature', in Gabriella Giannachi and Nigel Stewart (eds), *Performing Nature: Explorations in Ecology and the Arts* (Bern: Peter Lang), pp. 165–79

Aston, Elaine, 2015, 'Caryl Churchill's Dark Ecology', in Carl Lavery and Clare Finburgh- Delijani (eds), *Rethinking the Theatre of the Absurd: Ecology, the Environment and the Greening of the Modern Stage* (London: Bloomsbury), pp. 59–76

Austin, J. L., 1962, *How to Do Things with Words* (Oxford: Clarendon Press)

Badiou, Alain, 2009, 'L'Hypothèse communiste. Interview d'Alain Badiou par Pierre Gaultier', *Grand Soir*, 6 August, http://www.legrandsoir.info/L-hyptothese-communiste-interview-d-Alain-Badiou-par-Pierre.html (accessed 22 June 2022)

Balme, Christopher and Tracy C. Davis (eds), 2017, *A Cultural History of Theatre, Vols 1–6* (London: Bloomsbury)

Bamford, Kiff, 2012, *Lyotard and the Figural in Performance, Art, and Writing* (London: Bloomsbury)

Barad, Karen, 2003, 'Posthumanist Performativity: Towards an Understanding of How Matter Comes to Matter', *Signs: Journal of Women in Culture and Society*, 28:3, 801–31

Barad, Karen, 2007, *Meeting the Universe Halfway: Quantum Physics and the Entanglement of Matter and Meaning* (Durham, NC: Duke University Press)

Barad, Karen, 2010, 'Quantum Entanglements and Hauntological Relations of Inheritance: Dis/continuities, Spacetime Enfoldings and Justice to Come', *Derrida Today*, 3:2, 240–68

Barad, Karen, 2012 'On Touching the Inhuman That Therefore I Am', *Differences: A Journal of Feminist Cultural Studies*, 23:3, 206–33

Barad, Karen, 2014, 'Diffracting Diffraction: Cutting Together-Apart', *Parallax*, 20:3, 168–87

Barad, Karen, 2017, 'No Small Matter: Mushroom Clouds, Ecologies of Nothingness, and Strange Topologies of Spacetimemattering', in Anna Tsing, Nils Bubandt, Elaine Gan and Heather Anne Swanson (eds), *Arts of Living on a Damaged Planet* (Minneapolis: University of Minnesota Press), pp. G103–20

Barad, Karen, 2018, 'Troubling Time/s and Ecologies of Nothingness: Re-turning, Re-membering, and Facing the Incalculable', in Matthias Fritsch, Philippe Lynes and David Wood (eds), *Eco-Deconstruction: Derrida and Environmental Philosophy* (Fordham: Fordham University Press), pp. 206–48

Barba, Eugenio and Nicola Savarese, 1991, *A Dictionary of Theatre Anthropology*, trans. Richard Fowler (London: Centre for Performance Research/Routledge)

Barber, Stephen, 1993, *Antonin Artaud: Blows and Bombs* (London: Faber & Faber)

Barish, Jonas, 1981, *The Anti-Theatrical Prejudice* (Berkeley and Los Angeles: University of California Press)

Barthes, Roland, 1964 [1954], 'Le Théâtre de Baudelaire', in *Essais critiques* (Paris: Seuil), pp. 304–10

Bataille, Georges, 1988 [1949], *The Accursed Share: Vol. 1: Consumption*, trans. Robert Hurley (New York: Zone Books)

Bataille, Georges, 1992 [1973], *Theory of Religion*, trans. Robert Hurley (New York: Zone Books)

Bateson, Gregory, 1972, *Steps to an Ecology of Mind: Collected Essays in Anthropology, Psychiatry, Evolution and Epistemology* (Chicago: University of Chicago Press)

Bateson, Gregory, 2002, *Mind and Nature: A Necessary Unity* (Cresskill: Hampton Press)

Battista, Silvia, 2018, *Posthuman Spiritualities in Contemporary Performance: Politics, Ecologies and Perceptions* (Basingstoke: Palgrave Macmillan)

Baudrillard, Jean, 1996 [1990], *Cool Memories II*, trans. Chris Turner (Cambridge: Polity)

Bayley, Annouchka (ed.), 2020, *Performance Research*: 'On Diffraction', 25:5

Beer, Tanja, 2016, 'Ecomaterialism in Scenography', *Theatre and Performance Design*, 2:1–2, 161–72

Bell, Chris, 2020, 'Unsettling Existence: Land Acknowledgement in Contemporary Indigenous Performance', *Performance Research: On Dark Ecologies*, 25:2, 141–8

Bellamy Foster, John, 2000, *Marx's Ecology: Materialism and Nature* (New York: Monthly Review Press)

Benjamin, Walter, 1969, *Illuminations: Essays and Reflections*, trans. Harry Zorn (New York: Schocken Books)

Bennett, Jane, 2001 *The Enchantment of Modern Life: Attachments, Crossings, and Ethics* (Princeton: Princeton University Press)

Bennett, Jane, 2010, *Vibrant Matter: A Political Ecology of Things* (Durham, NC: Duke University Press)

Berardi, Franco, 2011, *After the Future*, trans. Arianna Bove (Edinburgh: AK Press)

Berardi, Franco, 2015, *Phenomenology of the End* (Cambridge, MA: MIT Press)

Bergthaller, Hannes, 2018, 'Beyond Ecological Crisis: Niklas Luhmann's Theory of Social System', *Electronic Book Review*, https://electronicbookreview.com/ebr-author/hannes-bergthaller/ (accessed 18 June 2021)

Bharucha, Rustom, 1978, 'Eclecticism, Oriental Theatre and Artaud', *Theater*, 9:3, 50–9

Bleeker, Maaike, 2020, 'Posthuman Landscapes', in Peter Eckersall and Kristof van Baarle (eds), *Machine Made Silence: The Art of Kris Verdonck* (Aberystwyth: Performance Research Books), pp. 92–100

Bogost, Ian, 2012, *Alien Phenomenology or What It's Like to Be a Thing* (Minneapolis: University of Minnesota Press)

Bonta, Mark and John Protevi, 2003, *Deleuze and Geophilosophy: A Guide and Glossary* (Edinburgh: Edinburgh University Press)

Borie, Monique, 1989, *Artaud: Le Théâtre et le retour aux sources: une approche anthropologique* (Paris: Gallimard)

Bottoms, Stephen, J., 2012, 'Climate Change Science on the London Stage', *Wiley Interdisciplinary Reviews: Climate Change*, 3:4, 339–48

Bottoms-Scott, Stephen, J., 2019, 'Holding Back the River', *Theatre Journal*, 71:4, E 61–7

Bottoms, Stephen J. and Matthew Goulish (eds), 2007, *Small Acts of Repair: Performance, Ecology and Goat Island* (London: Routledge)

Bradnock, Lucy, 2021, *No More Masterpieces: Modern Art after Artaud* (New Haven: Yale University Press)

Braidotti, Rosi, 2013, *The Posthuman* (Cambridge: Polity)
Braidotti, Rosi and Simone Bignall (eds), 2019, *Posthuman Ecologies: Complexity and Process after Deleuze* (London: Rowman and Littlefield)
Braidotti, Rosi and Maria Hlavajova (eds), 2018, *Posthuman Glossary* (London: Bloomsbury)
Brassier, Ray, 2007, *Nihil Unbound: Enlightenment and Extinction* (Basingstoke: Palgrave Macmillan)
Brecht, Bertolt, 1964 [1938], 'The Street Scene: A Basic Model for an Epic Theatre', in *Brecht on Theatre*, trans. John Willett (London: Methuen), pp. 121–9
Brecht, Bertolt, 1976, *Bertolt Brecht: Poems 1913–56*, ed. John Willett and Ralph Manheim, trans. John Willett *et al.* (London: Eyre Methuen)
Brown, Paul, 2009, 'Review Essay: Theatre, Ecological Sanity, and Finding Baz Kershaw', *Performance Paradigm, A Journal of Performance and Contemporary Culture*, 5:1, https://www.performanceparadigm.net/index.php/journal/article/view/72/7 (accessed 21 June 2021)
Brunel, Pierre, 1982, *Théâtre et cruauté ou Dionysos profané* (Paris: Librairie des Méridiens)
Burns, Elizabeth, 1972, *Theatricality* (London: Longman)
Butler, Judith, 1988, 'Performative Acts and Gender Constitution: An Essay in Phenomenology and Feminist Theory', *Theatre Journal*, 40:4, 519–31
Cage, John, 2019 [1965], *Diary: How to Improve the World (You Will Only Make Matters Worse)* (New York: Siglio Press)
Campbell, Alyson and David Farrier (eds), 2016, *Queer Dramaturgies: International Perspectives* (Basingstoke: Palgrave Macmillan)
Carasco, Raymonde, 2006, 'Approche de la pensée tarahumara', in Gauillaume Fau (ed.), *Antonin Artaud* (Paris: Gallimard), pp. 131–41
Carlson, Marvin, 2002, 'The Resistance to Theatricality', *Substance*, 31:2–3, 238–50
Carlson, Marvin, 2003, *The Haunted Stage: Theatre as Memory Machine* (Ann Arbor: University of Michigan Press)
Carson, Anne, 2006, *Decreation: Poetry, Essays, Opera* (New York: Vintage Books)
Carson, Rachel, 2000 [1962], *Silent Spring* (London: Penguin Books)
Carter, Paul, 2011, 'Archipelago: The Shape of the Future', *Antithesis*, 21, 11–24
Carter, Paul, 2019, *Decolonizing Governance: Archipelagic Thinking* (London: Routledge)
Caygill, Howard, 2015, 'Artaud-Immunity: Derrida and the *Mômo*', *Derrida Today*, 8:2, 113–35
Cervera, Felipe, 2019, 'Performance and Eschatological Politics', in Peter Eckersall and Helena Grehan (eds), *The Routledge Companion to Theatre and Politics* (London: Routledge), pp. 295–8
Chakrabarty, Dipesh, 2009, 'The Climate of History: Four Theses', *Critical Enquiry*, 35:2, 197–222
Chaudhuri, Una, 1994, '"There Must Be a Lot of Fish in that Lake": Toward an Ecological Theatre', *Theater*, 25:1, 23–31
Chaudhuri, Una, 1995, *Staging Place: The Geography of Modern Drama* (Ann Arbor: University of Michigan Press)
Chaudhuri, Una, 2016a, 'Anthropo-scenes: Staging Climate Chaos in the Drama of Bad Ideas', in Siân Adiseshiah and Louise LePage (eds), *Twenty-First Century Drama: What Happens Now* (Basingstoke: Palgrave Macmillan), pp. 303–21

Chaudhuri, Una, 2016b, *The Stage Lives of Animals: Zooësis and Performance* (London: Routledge)
Chaudhuri, Una and Shonni Enelow, 2013, *Research Theatre, Climate Change and the Ecocide Project* (Basingstoke: Palgrave Macmillan)
Chaudhuri, Una and Elinor Fuchs (eds), 2002, *Land/Scape/Theater* (Ann Arbor: University of Michigan Press)
Chen, Mel Y., 2011, 'Toxic Animacies: Inanimate Affections', *GLQ: A Journal of Lesbian and Gay Studies*, 17: 2–3, 265–86
Chen, Mel Y., 2012, *Animacies: Biopolitics, Racial Mattering, and Queer Affect* (Durham: Duke University Press)
Clark, Nigel, 2011, *Inhuman Nature: Sociable Life on a Dynamic Planet* (London: Sage)
Clark, Timothy (ed.), 2010, *Oxford Literary Review*: 'Deconstruction, Environmentalism, and Climate Change', 32:1
Clark, Timothy, 2015, *Ecocriticism on the Edge: The Anthropocene as a Threshold Concept* (London: Bloomsbury)
Clark, Timothy, 2019, *The Value of Ecocriticism* (Cambridge: Cambridge University Press)
Cless, Downing, 2010, *Ecology and Environment in Western Drama* (London: Routledge)
Clifford, James, 1988, *The Predicament of Culture: Twentieth-Century Ethnography, Literature, and Art* (Cambridge, MA: Harvard University Press)
Clifford, James, 2013, *Returns: Becoming Indigenous in the Twenty-First Century* (Cambridge, MA: Harvard University Press)
Cohen, Jeffrey Jerome (ed.), 1996, *Monster Theory: Reading Culture* (Minneapolis: University of Minnesota Press)
Cohen, Jeffrey Jerome, 2015, *Stone: An Ecology of the Inhuman* (Minneapolis: University of Minnesota Press)
Cohen, Jeffrey Jerome and Lowell Duckert (eds), 2017, *Veer Ecology: A Companion for Environmental Thinking* (Minneapolis: University of Minnesota Press)
Colebrook, Claire, 2006, *Deleuze: A Guide for the Perplexed* (London: Bloomsbury)
Colebrook, Claire, 2014, *The Death of the Posthuman: Essays on Extinction*, vol. 1 (Ann Arbor: Open Humanities Press)
Colebrook, Claire, 2015, 'Who Comes after the Post-Human', in Jon Roffe and Hannah Stark (eds), *Deleuze and the Non/Human* (Basingstoke: Palgrave Macmillan), pp. 215–34
Colebrook, Claire, 2018, 'Extinguishing Ability: How We Became Postextinction Person', in Matthias Fritsch, Philippe Lynes and David Wood (eds), *Eco-Deconstruction: Derrida and Environmental Philosophy* (Fordham: Fordham University Press), pp. 261–7
Corrieri, Augusto, 2016, *In Place of a Show: What Happens inside Theatres when Nothing Happens* (London: Bloomsbury)
Corrieri, Augusto, 2017, 'The Rock, the Butterfly, the Moon, and the Cloud: Notes on Dramaturgy in an Ecological Age', in Konstantina Georgelou, Efrosini Protopapa and Danae Theodoridou (eds), *The Practice of Dramaturgy: Working on Actions in Performance* (Amsterdam: Valiz), pp. 233–46
Corvin, Michel, 2014, *L'Homme en trop: l'abhumanisme dans le théâtre contemporain* (Besançon: Solitaires Intempestifs)

Craig, Edward Gordon, 1908, 'The Actor and the Über-Marionette', *The Mask*, 1:2, 1–15
Crary, Jonathan, 2022, *Scorched Earth: Beyond the Digital Age to a Post-Capitalist World* (London: Verso)
Cronon, William, 1996, *Uncommon Ground: Rethinking the Human Place in Nature* (New York: W. W. Norton & Company)
Cruickshank, Julie, 2005, *Do Glaciers Listen? Local Knowledge, Colonial Encounters, and Social Imagination* (Vancouver: University of British Columbia Press)
Crutzen, Paul J. and Eugene Stoermer, 2000, 'The Anthropocene', *IGBP Newsletter*, 41, 17–18
Cull, Laura, 2009, 'How Do You Make Yourself a Theatre Without Organs? Deleuze, Artaud and the Concept of Differential Presence', *Theatre Research International*, 34:3, 243–55
Cull, Laura, 2013, *Theatres of Immanence: Deleuze and the Ethics of Performance* (Basingstoke: Palgrave Macmillan)
Danowski, Déborah and Eduardo Viveiros de Castro, 2017 [2014], *The Ends of the World*, trans. Rodrigo Nunes (Cambridge: Polity)
Davis, Heather and Etienne Turpin (eds), 2015, *Art in the Anthropocene: Encounters among Aesthetics, Politics, Environments and Epistemologies* (London: Open Humanities Press)
Davis, Tracy C. and Thomas Postlewait (eds), 2003, *Theatricality* (Cambridge: Cambridge University Press)
De Julio, Maryann, 1997, 'Nancy Spero's *Codex Artaud*', *Dalhousie French Studies*, 39:40, 137–50
De La Cadena, Marisol, 2015, *Earth Beings: Ecologies of Practice across Andean Worlds* (Durham, NC: Duke University Press)
DeLanda, Manuel, 1997, *A Thousand Years of Nonlinear History* (New York: Zone Books)
Deleuze, Gilles, 1995 [1993], *Negotiations: 1972–1990*, trans. Martin Joughin (New York: Columbia University Press)
Deleuze, Gilles, 2003 [1981], *Francis Bacon: The Logic of Sensation*, trans. Daniel W. Smith (London: Continuum)
Deleuze, Gilles, 2004 [2002], *Desert Islands and Other Texts 1953–1974*, ed. David Lapoujade, trans. Michael Taormina (Cambridge, MA: MIT Press)
Deleuze, Gilles, 2008 [1964], *Proust and Signs*, trans. Richard Howarth (London: Continuum)
Deleuze, Gilles and Félix Guattari, 1983 [1972], *Anti-Oedipus, Capitalism and Schizophrenia*, trans. Robert Hurley, Mark Seem and Helen R. Lane (Minneapolis: University of Minnesota Press)
Deleuze, Gilles and Félix Guattari, 1987 [1980], *A Thousand Plateaus: Capitalism and Schizophrenia*, trans. Brian Massumi (Minneapolis: University of Minnesota Press)
Deleuze, Gilles and Félix Guattari, 1994 [1991], *What Is Philosophy?*, trans. Graham Burchell and Hugh Tomlinson (London: Verso)
Demaitre, Ann, 1972, 'The Theatre of Cruelty and Alchemy: Artaud and "Le Grand Œuvre"', *Journal of the History of Ideas*, 33:2, 237–50
Demos, T. J., 2016, *Decolonizing Nature: Contemporary Art and the Politics of Ecology* (Cambridge, MA: MIT Press)

Derrida, Jacques, 1974 [1967], *Of Grammatalogy*, trans. Gayatri Chakravorty Spivak (Baltimore: Johns Hopkins University Press)
Derrida, Jacques, 1978 [1967], *Writing and Difference*, trans. Alan Bass (London: Routledge)
Derrida, Jacques, 1981 [1968], *Dissemination*, trans. Barbara Johnson (Chicago: University of Chicago Press)
Derrida, Jacques, 1982 [1972], *Margins of Philosophy*, trans. Alan Bass (Chicago: University of Chicago Press)
Derrida, Jacques, 1984, '"No Apocalypse, Not Now" (Full Speed Ahead, Seven Missiles, Seven Missiles)', *Diacritics*, 14:2, 20–31
Derrida, Jacques, 1991. '"Eating Well": Or the Calculation of the Subject: An Interview with Jacques Derrida', in E. Cadava, and J.-L. Nancy (eds), *Who Comes after the Subject* (London: Routledge), pp. 96–119
Derrida, Jacques, 1994 [1992], *Specters of Marx: The State of the Debt, the Work of Mourning and the New International*, trans. Peggy Kamuf (London: Routledge)
Derrida, Jacques, 1997 [1994], *The Politics of Friendship*, trans. George Collins (London: Verso)
Derrida, Jacques, 2007 [2005], *Learning to Live Finally: The Last Interview*, trans. Pascale-Anne Brault and Michael Naas (Basingstoke: Palgrave Macmillan)
Derrida, Jacques, 2008 [2006], *The Animal that Therefore I Am*, trans. David Wills, (Fordham: Fordham University Press)
Derrida, Jacques, 2009 [2008], *The Beast and the Sovereign, Volume 1*, trans. Geoffrey Bennington (Chicago: University of Chicago Press)
Derrida, Jacques, 2017 [2002], *Artaud the Moma: Interjections of Appeal*, trans. Peggy Kamuf (New York: Columbia University Press)
Derrida, Jacques and Paule Thévenin, 1996, *The Secret Art of Antonin Artaud*, trans. Mary Ann Caws (Cambridge, MA: MIT Press)
Dodds, Joseph, 2011, *Psychoanalysis and Ecology at the Edge of Chaos: Complexity Theory, Deleuze/Guattari and Psychoanalysis for a Climate in Crisis* (London: Routledge)
Dolan, Jill, 2005, *Utopia in Performance: Finding Hope at the Theater* (Ann Arbor: University of Michigan Press)
Dolphijn, Rick and Iris van der Tuin (eds), 2012, *New Materialisms: Interviews and Cartographies* (Ann Arbor: Open Humanities Press)
Domínguez, Patricia, 2022, *Vegetal Matrix 2021* (London: Wellcome Trust)
Donald, Minty, 2014, 'Entided, Enwatered, Endwinded: Human/More-Than-Human Agencies in Site-Specific Performance', in Maria Schweitzer and Joanne Zerdy (eds), *Performing Objects and Theatrical Things* (Basingstoke: Palgrave Macmillan), pp. 118–31
Donald, Minty, 2019, 'Guddling About: An Ecological Practice with Water and other NonHuman Collaborators', *Geohumanities*, 5:2, 591–619
Donald, Minty and Nick Millar, 2014–16, *Guddling About*, https://guddlingabout.com/about/ (accessed 21 June 2022)
Eckersall, Peter, Helena Grehan and Edward Scheer, 2017, *New Media Dramaturgy: Performance, Media and New Materialism* (Basingstoke: Palgrave Macmillan)
Eckersall, Peter and Kristof van Baarle (eds), 2020, *Machine Made Silence: The Art of Kris Verdonck* (Aberystwyth: Performance Research Books)
Edelman, Lee, 2004, *No Future: Queer Theory and the Death Drive* (Durham, NC: Duke University Press)

Egan, Gabriel, 2006, *Green Shakespeare from Ecopolitics to Ecocriticism* (London: Routledge)
Elias, Amy J. and Christian Moraru (eds), 2015, *The Planetary Turn: Relationality and Geoaesthetics in the Twenty-First Century* (Evanston: Northwestern University Press)
Ellsworth, Elizabeth and Jamie Kruse (eds), 2012, *Making the Geologic Now: Response to Material Conditions of Contemporary Life* (New York: Punctum Books)
Epstein, Jean, 1988, 'Magnification', in Richard Abel (ed.), *French Film Theory and Criticism, A History/Anthology, 1907–39, Vol. 1* (Princeton: Princeton University Press), pp. 235–40
Escobar, Arturo, 2018, *Designs for the Pluriverse: Racial Interdependencies, Autonomy and the Making of Worlds* (Durham, NC: Duke University Press)
Esslin, Martin, 1976, *Antonin Artaud: The Man and His Work* (London: John Calder)
Estok, Simon, 2018, *The Ecophobia Hypothesis* (London: Routledge)
Etchells, Tim, 1997, 'Repeat Forever', in Adrian Heathfield, Fiona Templeton and Andrew Quick (eds), *Shattered Anatomies: Traces of the Body in Performance* (Bristol: Arnolfini Live), n.p.
Etchells, Tim, 1999, *Certain Fragments: Contemporary Performance and Forced Entertainment* (London: Routledge)
Evink, Eddo, 2020, 'Différance as Temporization and Its Problems', *International Journal of Philosophical Studies*, 28:3, 233–451
Fensham, Rachel, 2009, *To Watch Theatre: Essays on Genre and Corporeality* (Brussels: Peter Lang)
Fensham, Rachel, 2020, 'Ecological Combustion: The Atmospherics of the Bushfire as Choreography', *Ambiances* 6: 'Staging Atmospheres: Theatre and the Atmospheric Turn', – vol. 1, https://journals.openedition.org/ambiances/ (accessed 21 June 2021)
Fensham, Rachel, Eddie Paterson and Paul Rae (eds), 2018, *Performance Research*: 'On Climates', 23:3
Féral, Josette, 1997 [1982], 'Performance and Theatricality: The Subject Demystified', in Timothy Murray (ed.), *The Politics of Theatricality in Contemporary French Thought*, trans. Terese Lyons (Ann Arbor: University of Michigan Press), pp. 289–300
Féral, Josette, 2002, 'Theatricality: The Specificity of Theatrical Language', trans. Ronald P. Bermingham, *Substance*, 31:2–3, 94–108
Ferdinand, Malcom, 2022 [2019], *Decolonial Ecology: Thinking from the Caribbean World*, trans. Antony Paul Smith (Cambridge: Polity Press)
Finter, Helga, 1997, 'Antonin Artaud and the Impossible Theatre: The Legacy of the Theatre of Cruelty', *The Drama Review*, 41:4, 15–40
Fischer-Lichte, Erika, 1995, 'Theatricality: A Key Concept in Theatre and Cultural Studies', *Theatre Research International*, 20:2, 85–9
Fischer-Lichte, Erika, 2008 [2004], *The Transformative Power of Performance: A New Aesthetics*, trans. Saskya Iris Jain (London: Routledge)
Florêncio, João, 2014 *Strange Encounters: Performance in the Anthropocene* (PhD dissertation, Goldsmiths, University of London)
Florêncio, João, 2015, 'Enmeshed Bodies, Impossible Touch: The Object-Orientated World of Pina Bausch's *Café Müller*', *Performance Research*: 'On Anthropomorphism', 20:2, 53–9

Foucault, Michel, 1999 [1966], *The Order of Things: An Archaeology of the Human Sciences*, trans. Alan Sheridan (London: Routledge)
Fragkou, Marissia, 2019, *Ecologies of Precarity in the Twenty-First Century Theatre* (London: Bloomsbury)
Frémeaux, Isabelle and John Jordan, 2021, *We Are Nature Defending Itself: Entangling, Art, Activism, and Autonomous Zones* (London: Pluto)
Fried, Michael, 1967, 'Art and Objecthood', *Art Forum*, 5:10, 12–23
Fritsch, Matthias, Philippe Lynes and David Wood, 2018, 'Introduction', in Matthias Fritsch, Philippe Lynes and David Wood (eds), *Eco-Deconstruction: Derrida and Environmental Philosophy* (Fordham: Fordham University Press), pp. 1–26
Frost, Samantha, 2011, 'The Implications of the New Materialisms for Feminist Epistemology', in Heidi E. Grasswick (ed.), *Feminist Epistemology and the Philosophy of Science* (London: Springer), pp. 69–83
Fuchs, Elinor, 1994, 'Play as Landscape: Another Version of Pastoral', *Modern Drama*, 25:1, 44–51
Fukuyama, Francis, 2002, *Our Posthuman Future: Consequences of the Biotechnology Revolution* (New York: Farrar, Straus and Giroux)
Garner, Stanton B., 2006, 'Artaud, Germ Theory, and the Theory of Contagion', *Theatre Journal*, 58:1, 1–14
Garrard, Greg, 2012, *Ecocriticism*, 2nd edn (London: Routledge)
Gasché, Rodolphe, 2014, *Geophilosophy: On Gilles Deleuze and Félix Guattari's What Is Philosophy?* (Chicago: Northwestern University Press)
Genet, Jean, 1986, *Un capif amoureux* (Paris: Gallimard)
Genosko, Gary, 2000, 'The Life and Work of Félix Guattari: From Transversality to Ecosophy', in *The Three Ecologies*, trans. Ian Pindar and Paul Sutton (London: Continuum), pp. 46–78
Giannachi, Gabriella and Nigel Stewart (eds), 2005, *Performing Nature: Explorations in Ecology and the Arts* (Bern: Peter Lang)
Gibson, J. J., 1979, *The Ecological Approach to Visual Perception* (Boston: Houghton Miffin)
Gibson, William, 1984, *Neuromancer* (London: Gollancz)
Gil, José, 1998 [1985], *Metamorphoses of the Body*, trans. Stephen Muecke (Minneapolis: University of Minnesota Press)
Gilbert, Helen, 2013, 'Indigeneity, Time and the Cosmopolitics of Postcolonial Belonging in the Atomic Age', *Interventions*, 15:2, 195–210
Glissant, Édouard, 1997 [1990], *Poetics of Relation*, trans. Besty Wing (Ann Arbor: University of Michigan Press)
Goodall, Jane, 1990, 'The Plague and Its Powers in Artaudian Theatre', *Modern Drama*, 33: 529–42
Goodall, Jane, 1994, *Artaud and the Gnostic Drama* (Oxford: Clarendon Press)
Goodall, Jane, 2002, *Performance and Evolution in The Age of Darwin* (London: Routledge)
Gotman, Kélina, 2015, 'Exceptionalism, Schizophrenia, Artaud: On Judgment', *Performance Philosophy*, 1:1, 119–25
Gotman, Kélina, 2018, *Essays on Theatre and Change: Towards a Poetics of* (London: Routledge)
Greene, Naomi, 1994, '"All the Great Myths are Dark": Artaud and Fascism', in Gene A. Plunka (ed.), *Antonin Artaud and the Modern Theater* (Rutherford: Fairleigh Dickinson University Press), pp. 102–16

Grehan, Helena, 2018, 'First Nations Politics in a Climate of Refusal: Speaking and Listening but Failing to Hear', *Performance Research*: 'On Climates', 23:3, 7–13

Grosz, Elizabeth, 2008, *Chaos, Territory, Art: Deleuze and the Framing of the Earth* (New York: Columbia University Press)

Grosz, Elizabeth, 2017, *The Incorporeal: Ontology, Ethics and the Limits of Materialism* (New York: Columbia University Press)

Guattari, Félix, 1995 [1992], *Chaosmosis: An Ethico-Aesthetic Paradigm*, trans. Paul Bains and Julian Pefanis (Sydney: Power Publication)

Guattari, Félix, 2000 [1989], *The Three Ecologies*, trans. Ian Pindar and Paul Sutton (London: Continuum)

Guattari, Félix, 2015 [1972], *Psychoanalysis and Transversality: Texts and Interviews 1955–1971*, trans. Ames Hodges (Pasadena: Semiotext(e))

Hamilton, Jennifer Mae and Astrida Neimanis, 2018, 'Composting Feminisms and Environmental Humanities', *Environmental Humanities*, 10:2, 501–27

Haraway, Donna, 1985, 'A Manifesto for Cyborgs: Science, Technology and Socialist Feminism in the 1980s', *Socialist Review*, 80, 65–108

Haraway, Donna, 1991, *Simians, Cyborgs and Women: The Reinvention of Nature* (London: Routledge)

Haraway, Donna, 2003, *The Companion Species Manifesto: Dogs, People and Significant Otherness* (Chicago: Prickly Paradigm Press)

Haraway, Donna, 2008, *When Species Meet* (Minneapolis: University of Minnesota Press)

Haraway, Donna, 2015, 'Anthropocene, Capitalocene, Plantationocene, Chthulucene: Making Kin', *Environmental Humanities*, 6:1, 159–65

Haraway, Donna, 2016, *Staying with the Trouble: Making Kin in the Chthulucene* (Durham, NC: Duke University Press)

Haraway, Donna and Anna Tsing, 2019, 'Reflections on the Plantationocene: A Conversation with Donna Haraway and Anna Tsing', https://edgeeffects.net/haraway-tsing-plantationocene/ (accessed 22 June 2021)

Harney, Stefano and Fred Moten, 2013, *The Undercommons: Fugitive Planning & Black Study* (Wivenhoe: Minor Composition)

Harries, Martin, 2020, 'Double Take: *Theatre & Everyday Life: An Ethics of Performance* by Alan Read', *Theatre Research International*, 45:3, 362–4

Harris, Anne M. and Stacy Holman Jones, 2019, *The Queer Life of Things: Performance, Affect and the More-Than-Human* (Lanham: Lexington Books)

Harth, Dietrich, 2004, 'Artaud's Holy Theatre: A Case for Questioning the Relations between Ritual and Stage Performance', in Jens Kreinath (ed.), *The Dynamics of Changing Rituals: The Transformation of Religious Rituals within Their Social and Cultural Context* (New York: Peter Lang), pp. 73–85

Hartnell, Laura, 2020, 'Notes from the Semiotic Chora: Theatre and Performance as Affective Diffraction', *Performance Research*: 'On Diffraction', 25:5, 4–9

Hassan, Ihab, 1977, 'Prometheus as Performer: Towards a Posthumanist Culture?', *The Georgia Review*, 31:4, 830–50

Hayles, Katherine, 1999, *How We Became Posthuman: Virtual Bodies in Cybernetics, Literature, and Informatics* (Chicago: University of Chicago Press)

Heathfield, Adrian, 2018, 'Before Judson and Some Other Things', in Ana Janevski and Thomas J. Lax (eds), *Judson Dance Theatre: The Work Is Never Done* (New York: MoMA), pp. 36–43

Heathfield, Adrian, Fiona Templeton and Andrew Quick (eds), 1997, *Shattered Anatomies: Traces of the Body in Performance* (Bristol: Arnolfini Live)

Heddon, Dee and Sally Mackey, 2012, 'Environmentalism, Performance, and Applications: Uncertainties and Emancipations', *Research in Drama Education: The Journal of Applied Theatre and Performance*, 17:2, 163–92

Heidegger, Martin, 1971a [1936], 'The Origin of the Work of Art', in *Poetry, Language, Thought*, trans. Albert Hofstadter (New York: Harper Colophon), pp. 15–88

Heidegger, Martin, 1971b [1954], 'Building Dwelling Thinking', in *Poetry, Language, Thought*, trans. Albert Hofstadter (New York: Harper Colophon), pp. 143–62

Heidegger, Martin, 1977, *The Question Concerning Technology and Other Essays*, trans. William Lovitt (New York and London: Garland)

Heidegger, Martin, 2014 [1953], *Introduction to Metaphysics*, revised and expanded 2nd edn, trans. Gregory Fried and Richard Polt (New Haven: Yale University Press)

Heim, Wallace, 2012a, 'Epilogue: Thinking Forward', in Wendy Arons and Theresa J. May (eds), *Readings in Performance and Ecology* (Basingstoke: Palgrave Macmillan), pp. 211–16

Heim, Wallace, 2012b, 'Can a Place Learn', *Performance Research*: 'On Ecology', 17:4, 120–7

Heise, Ursula, 2008, *Sense of Place and Sense of Planet: The Environmental Imagination of the Global* (Oxford: Oxford University Press)

Herbrechter, Stefan, 2013, *Posthumanism: A Critical Analysis* (London: Bloomsbury)

Herzogenrath, Bernd (ed.), 2008, *An [Un]likely Alliance: Thinking Environment[s] with Deleuze/Guattari* (Newcastle-upon-Tyne: Cambridge Scholars)

Herzogenrath, Bernd (ed.), 2009, *Deleuze/Guattari & Ecology* (Basingstoke: Palgrave Macmillan)

Hilevaara, Katja and Emily Orley (eds), 2018, *The Creative Critic: Writing as/about Critic* (London: Routledge)

Hiltner, Ken (ed.), 2015, *Ecocriticism: The Essential Reader* (London: Routledge)

Hird, Myra J. and Kathryn Yusoff, 2019, 'Lines of Shite: Microbial-Mineral Chatter in the Anthropocene', in Rosa Braidotti and Simone Bignall (eds), *Posthuman Ecologies and Deleuzian Philosophy* (London: Rowman and Littlefield), pp. 265–82

Hudson, Julie, 2020, *The Environment on Stage: Scenery or Shapeshifter?* (London: Routledge)

Hughes, Jenny, 2009, 'Baz Kershaw: *Theatre Ecology: Environments and Performance Events*', *New Theatre Quarterly*, 25:1, 102–3

Hui, Yuk, 2020a, *Art and Cosmotechnics* (Minneapolis: University of Minnesota Press)

Hui, Yuk, 2020b, 'For a Planetary Thinking', *E-Flux Journal*, 114, https://www.e-flux.com/journal/114/366703/for-a-planetary-thinking/ (accessed 22 June 2022)

Infante, Manuela, 2020, 'Plant-Based Dramaturgy', *Transmission in Motion Seminar Series*, University of Utrecht

Ingold, Tim, 2012, 'Towards an Ecology of Materials', *The Annual Review of Anthropology*, 41, 427–42

Innes, Christopher, 1993, *Avant-Garde Theatre 1892–1992* (London: Routledge)

Iovino, Serenella and Serpil Oppermann (eds), 2014, *Material Ecocriticism* (Bloomington: Indiana University Press)
Jackson, Shannon, 2011, *Performing Art: Supporting Public* (London: Routledge)
Jameson, Fredric, 1992, *Postmodernism, or the Cultural Logic of Postmodernism* (London: Verso)
Jannarone, Kimberly, 2010, *Artaud and His Doubles* (Ann Arbor: University of Michigan Press)
Johnson, Dominic, 2008, 'Theatre, Intimacy, & Engagement: The Last Human Venue', *Contemporary Theatre Review*, 18:4, 514–15
Kaya, Hannah, 2020, 'Choreographies of Mourning', *Performance Research*: 'On Dark Ecologies', 25:2, 53–60
Kelleher, Joe, 2008, 'Theatre, Intimacy & Engagement: The Last Human Venue', *The Drama Review*, 54:2, 181–3
Kelleher, Joe, 2009, *Theatre & Politics* (Basingstoke: Palgrave Macmillan)
Kelleher, Joe, 2015a, *The Illuminated Theatre: Studies on the Suffering of Images* (London: Routledge)
Kelleher, Joe, 2015b, 'Recycling Beckett', in Carl Lavery and Clare Finburgh Delijani (eds), *Rethinking the Theatre of the Absurd: Ecology, the Environment and the Greening of the Modern Stage* (London: Bloomsbury), pp. 127–46
Kermode, Frank, 2000 [1967], *The Sense of an Ending: Studies in the Theory of Fiction*, 2nd edn (Oxford: Oxford University Press)
Kershaw, Baz, 2007, *Theatre Ecology: Environments and Performance Events* (Cambridge: Cambridge University Press)
Kershaw, Baz, 2012a, 'Dancing with Monkeys? On Performance Commons and Scientific Experiments', in Wendy Arons and Theresa J. May (eds), *Readings in Performance and Ecology* (Basingstoke: Palgrave Macmillan), pp. 59–76
Kershaw, Baz, 2012b, 'This is the way the world ends, not ...? On Performance Compulsion and Climate Change', *Performance Research*: 'On Ecology', 17:4, 5–17
Kershaw, Baz, 2016, 'Projecting Climate Scenarios, Landscaping Nature and Knowing Performance: On Becoming Performed by Ecology', *Green Letters: Studies in Ecocriticism, Performance and Ecology: What Can Theatre Do?*, 20:3, 270–89
Kirby, Vicki, 2011, *Quantum Anthropologies: Life at Large* (Durham, NC: Duke University Press)
Kirby, Vicki (ed.), 2017, *What If Culture Was Nature All Along?* (Edinburgh: Edinburgh University Press)
Kirby, Vicki, 2018, 'Un/Limited Ecologies', in Matthias Fritsch, Philippe Lynes and David Wood (eds), *Eco-Deconstruction: Derrida and Environmental Philosophy* (Fordham: Fordham University Press), pp. 121–40
Kleist, Heinrich von, 1972 [1810], 'On the Marionette Theatre', trans. Thomas G. Neumiller, *The Drama Review*, 16:3, 22–6
Kramer, Paula, 2012, 'Bodies, Rivers, Rocks and Trees: Meeting Agentic Materiality in Contemporary Outdoor Dance Practices', *Performance Research*: 'On Ecology', 17:4, 83–91
Kurzweil, Ray, 2005, *The Singularity Is Near: When Humans Transcend Biology* (New York: Penguin)

Lacan, Jacques, 1972, 'Seminar on the *Purloined Letter*', trans. Jeffrey Mehlman, *Yale French Studies*, 48: 39–72

Lapworth, Andrew, 2020, 'Gilbert Simondon and the Technical Mentalities and Transindividual Affects of Art-Science', *Body & Society*, 26:1, 107–34

Latour, Bruno, 1993 [1991], *We Have Never Been Modern*, trans. Catherine Porter (Cambridge, MA: Harvard University Press)

Latour, Bruno, 2004 [1999], *Politics of Nature: How to Bring the Sciences into Democracy*, trans, Catherine Porter (Cambridge, MA: Harvard University Press)

Latour, Bruno, 2015, *Face à Gaïa: Huit conférences sur le nouveau régime climatique* (Paris: La Découverte)

Latour, Bruno, 2017 [2015], *Facing Gaia: Eight Lectures on the New Climatic Regime*, trans. Catherine Porter (Cambridge: Polity)

Latour, Bruno, 2018 [2017], *Down to Earth: Politics in the New Climatic Regime*, trans. Catherine Porter (Cambridge: Polity)

Latour, Bruno and Peter Weibel (eds), 2020, *Critical Zones: The Science and Politics of Landing on Earth* (Cambridge, MA: MIT Press)

Lavery, Carl, 2009, 'Mourning Walk and Pedestrian Performance: History, Aesthetics and Ethics', in Roberta Mock (ed.), *Walking, Writing, and Performance: Autobiographical Texts by Deirdre Heddon, Carl Lavery, and Phil Smith* (Bristol: Intellect), pp. 41–56

Lavery, Carl, 2013, 'The Ecology of the Image: The Environmental Politics of Philippe Quesne and Vivarium Studio', *French Cultural Studies*, 24:3, 264–78

Lavery, Carl, 2015, '*Performing Paris: An Ecography of Meridians and Atmospheres*', in Nicolas Whybrow (ed.), *Performing Cities* (Basingstoke: Palgrave Macmillan), pp. 56–79

Lavery, Carl, 2016a, 'Late Modernism', in Maggie Gale and John F. Deeney (eds), *The Routledge Drama Anthology: From Modernism to Contemporary Performance* (London and New York: Routledge), pp. 546–67

Lavery, Carl, 2016b, 'Introduction: Performance and Ecology: What Can Theatre Do?', *Green Letters: Studies in Ecocriticism*, 20:3: 229–36

Lavery, Carl, 2019a, 'Comment penser l'image écologique dans le théâtre contemporain', *Théâtre*, 4, https://www.theatre.com/2019/03/29/comment-penser-limage-ecologique/ (accessed 22 June 2021)

Lavery, Carl, 2019b, 'Animating Tangible Futures', *Performance Research*: 'On Animism', 24:6, 29–37

Lavery, Carl, 2020, 'Thinking Materially: Verdonck in the Anthropocene', in Peter Eckersall and Kristof van Baarle (eds), *Machine Made Silence: The Art of Kris Verdonck* (Aberystwyth: Centre for Performance Research), pp. 68–97

Lavery, Carl and Clare Finburgh-Delijani (eds), 2015, *Rethinking the Theatre of the Absurd: Ecology, Environment, and the Greening of the Modern Stage* (London: Bloomsbury)

Lehmann, Hans-Thies, 2006 [1999], *Postdramatic Theatre*, trans. Karen Jürs-Munby (London: Routledge)

Lehmann, Hans-Thies, 2013, 'A Future for Tragedy? Remarks on the Political and Postdramatic', in Karen Jürs-Munby, Jerome Carroll and Steve Giles (eds), *Postdramatic Theatre and the Political: International Perspectives on Contemporary Performance* (London: Bloomsbury), pp. 87–110

Lehmann, Hans-Thies, 2016, *Tragedy and the Dramatic Theatre*, trans. Erik Butler (London and New York: Routledge)

LeMenager, Stephanie, 2014, *Living Oil: Petroleum Culture in the American Century* (Oxford: Oxford University Press)
Levine, Gabriel, 2020, 'Black-Light Ecologies: Punctuate! Theatre's *Bears* Wipes Off the Oil', *Performance Research*: 'On Dark Ecologies', 25:2, 45–52
Lévi-Strauss, Claude, 1955, *Tristes tropiques* (Paris: Plon)
Lewis, Simon and Mark Maslin, 2015, 'Defining the Anthropocene', *Nature*, 519, 171–80
Lilburn, Tim, 2019, *Living in the World: As If It Were Home* (Newcastleton: Xylem Books)
Lindqvist, Sven, 1992, *Exterminate All the Brutes* (Cambridge: Granta)
Llana, Jazmin Badong, 2018, 'A Sea of Stories: Archipelagic Gatherings and RoRo Journey', *Performance Research*: 'On Centenaries', 23:4–5, 256–61
Lucie, Sarah, 2020, *Acting Objects: Staging New Materialisms, Posthumanism and the Ecocritical Crisis in Contemporary Performance* (PhD dissertation, City University of New York)
Luckhurst, Mary and Emilie Morin (eds), 2015, *Theatre and Ghosts: Materiality, Performance and Modernity* (Basingstoke: Palgrave Macmillan)
Luhmann, Niklas, 1986 [1989], *Ecological Communication*, trans. John Bednarz (Chicago: University of Chicago Press)
Lynes, Philippe, 2018, 'The Posthuman Promise of the Earth', in Matthias Fritsch, Philippe Lynes and David Wood (eds), *Eco-Deconstruction: Derrida and Environmental Philosophy* (Fordham: Fordham University Press), pp. 101–20
Lyotard, Jean-François, 1971, *Discours, figure* (Paris: Klincksiek)
Lyotard, Jean-François, 1988, *Peregrinations: Law, Form, Event* (New York: Columbia University Press)
Lyotard, Jean-François, 1991 [1989], *The Inhuman: Reflections on Time*, trans. Geoffrey Bennington and Rachel Bowlby (Cambridge: Polity)
Lyotard, Jean-François, 1993a [1989], '*Oikos*', in Bill Readings and Kevin Paul Geiman (eds), *Jean François Lyotard: Political Writings* (London: UCL Press), pp. 96–107
Lyotard, Jean-François, 1993b, 'Gesture and Commentary', *The Jerusalem Philosophical Quarterly*, 42, 37–48
Lyotard, Jean-François, 1997a [1976], 'The Tooth, the Palm', trans. Anne Knabb and Michel Benamou, in Timothy Murray (ed.), *Mimesis, Masochism, and Mime: The Politics of Theatricality in Contemporary French Thought* (Ann Arbor: University of Michigan Press), pp. 282–8
Lyotard, Jean-François, 1997b [1993] *Postmodern Fables*, trans. Georges Van Den Abbeele (Minneapolis: University of Minnesota Press)
Lyotard, Jean-François, 1999, 'Anamnesis of the Visible 2', trans. John Ronan, *Qui Parle*, 11:2: 21–36
Lyotard, Jean-François, 2009a [1998], *Karel Appel: Un geste de couleur / Karel Appel: A Gesture of Colour*, trans. Vlad Ionescu and Peter W. Milne (Leuven: Leuven University Press)
Lyotard, Jean-François, 2009b, *Enthusiasm: The Kantian Critique of History*, trans. Georges Van Den Abbeele (Stanford: Stanford University Press)
MacDonald, Graeme, 2015, 'Commentary', in Graeme MacDonald (ed.), *The Cheviot, the Stag and the Black, Black Oil* (London: Bloomsbury), pp. 17–68
Mallarmé, Stéphane, 2006, *Collected Poems and Other Verse*, trans. E. H. and A. M. Blackmore (Oxford: Oxford University Press)

Mallarmé, Stéphane, 2018 [1957], *The Book*, trans. Sylvia Gorelick (Cambridge: Exact Change)
Manning, Erin, 2006, *Politics of Touch: Senses, Movement, Sovereignty* (Minneapolis: University of Minnesota Press)
Manning, Erin, 2013, *Always More Than One: Individuation's Dance* (Durham, NC: Duke University Press)
Manning, Erin, 2016, *The Minor Gesture* (Durham, NC: Duke University Press)
Manuel, Pedro, 2016, *Theatre without Actors: Rehearsing New Modes of Co-Presence* (PhD dissertation, University of Utrecht)
Marder, Michael, 2018, 'Ecology as Event', in Matthias Fritsch, Philippe Lynes and David Wood (eds), *Eco-Deconstruction: Derrida and Environmental Philosophy* (Fordham: Fordham University Press), pp. 141–64
Marranca, Bonnie, 1977, *The Theatre of Images* (New York: PAJ)
Marranca, Bonnie, 1996, *Ecologies of Theater: Essays at the Century Turning* (Baltimore: Johns Hopkins University Press)
Mauss, Marcel, 1973 [1935], 'Techniques of the Body', *Economy and Society*, 2:1, 70:88
May, Theresa J., 2010, 'Kneading Marie Clements' Burning Vision', *Canadian Theatre Review*, 144, 5–12
May, Theresa J., 2020, *Earth Matters on Stage: Ecology and Environment in American Theatre* (London and New York: Routledge)
Mbembe, Achille, 2019 [2016], *Necropolitics*, trans. Steven Corcoran (Durham, NC: Duke University Press)
McKenzie, Jon, 2001, *Perform or Else: From Discipline to Performance* (London: Routledge)
McKinney, Joslin, 2015, 'Scenographic Materialism: Affordance and Extended Cognition in Kris Verdonck's Actor#1', *Theatre and Performance Design*, 1:1–2, 79–93
McKinney, Joslin and Scott Palmer, 2017, *Scenography Expanded: An Introduction to Contemporary Performance Design* (London: Bloomsbury).
McKittrick, Katherine, 2013, 'Plantation Futures', *Small Axe*, 17:3, 1–15
McTighe, Trish, 2017, 'Museum, Furniture, Men: The Queer Ecology of *I Am My Own Wife*', *Modern Drama*, 60:2, 150–68
Meillassoux, Quentin, 2008, *After Finitude: An Essay on the Necessity of Contingency*, trans. Ray Brassier (London: Continuum)
Miles, Malcolm, 2014, *Eco-Aesthetics: Art, Literature and Architecture in a Period of Climate Change* (London: Bloomsbury)
Montag, Daro (ed.), 2008, *Artful Ecologies – Art, Nature, and Environment Conference* (Falmouth: Festerman)
Moore, Jason W., 2013, *Capitalism in the Web of Life: Ecology and the Accumulation of Capital* (London: Verso)
Moore, Jason W., (ed.), 2016, *Anthropocene or Capitalocene? Nature, History, and the Crisis of Capitalism* (Oakland: PM Press)
Moravec, Hans, 1988, *Mind Children: The Future of Robot and Human Intelligence* (Cambridge, MA: Harvard University Press)
Mortimer-Sandilands, Catriona, 2008, 'Landscape, Memory and Forgetting: Thinking Through (My Mother's) Body and Place', in Stacy Alaimo and Susan Hekman (eds), *Material Feminisms* (Bloomington: Indiana University Press), pp. 265–87

Morton, Timothy, 2007, *Ecology without Nature: Rethinking Environmental Ethics* (New Haven: Yale University Press)
Morton, Timothy, 2010a, *The Ecological Thought* (Cambridge, MA: Harvard University Press)
Morton, Timothy, 2010b, 'Queer Ecology', *PMLA*, 125:2, 272–83
Moten, Fred, 2003, *In the Break: The Aesthetics of Black Radical Tradition* (Minneapolis: University of Minnesota Press)
Murphy, Jay, 2016, *Artaud's Metamorphoses: From Hieroglyphs to Bodies without Organs* (New York: Pavement Books)
Murphy, Jay, 2021, *New Media and the Artaud Effect* (Cham: Palgrave Macmillan)
Murray, Ros, 2014, *Antonin Artaud: The Scum of the Soul* (Basingstoke: Palgrave Macmillan)
Murray, Timothy (ed.), 1997, 'Introduction: The Mise-en-Scene of the Cultural', in *Mimesis, Masochism, and Mime: The Politics of Theatricality in Contemporary French Thought* (Ann Arbor: University of Michigan Press), pp.1–26
Murris, Karin and Vivienne Bozalek, 2019, 'Diffraction and Response-able Reading of Texts: The Relational Ontologies of Barad and Deleuze', *International Journal of Qualitative Studies in Education*, 32:7, 872–86
Naas, Michael, 2018, 'E-Phemera: Of Deconstruction, Biodegradability, and Nuclear War', in Matthias Fritsch, Philippe Lynes and David Wood (eds), *Eco-Deconstruction: Derrida and Environmental Philosophy* (Fordham: Fordham University Press), pp. 197–205
Nail, Thomas, 2021, *Theory of the Earth* (Stanford: Stanford University Press)
Nancy, Jean-Luc, 1997 [1993], *The Gravity of Thought*, trans. François Raffoul and Gregory Recco (Atlantic Highlands: Humanities Press International)
Nancy, Jean-Luc, 2014 [2012], *After Fukushima: The Equivalence of Catastrophes*, trans. Charlotte Mandell (Fordham: Fordham University Press)
National Ocean Service, 2024, https://oceanservice.noaa.gov/facts/archipelago.html (accessed 19 August 2024)
Nayar, Pramod K., 2014, *Posthumanism* (Cambridge: Polity)
Nelhaus, Tobin, Bruce McConachie, Carol Fisher Sorgenfrei and Tamara Underinder, 2016, *Theatre Histories: An Introduction*, 2nd edn (London: Routledge)
Neve, Christopher, 2021, *Unquiet Landscape: Places and Ideas in the 20th-Century British Painting*, new edn (London: Thames&Hudson)
Neyrat, Frédéric, 2009, *Instructions pour une prise d'âmes: Artaud et l'envoûtement occidental* (Strasbourg: La Phocide)
Nicholson-Sanz, Michelle, 2020, 'The Performance of Water Governance as Cultural Heritage in Peru', *Contemporary Theatre Review*, 30:4, 509–24
Nixon, Rob, 2013, *Slow Violence and the Environmentalism of the Poor* (Cambridge, MA: Harvard University Press)
Norman, Sally Jane, 2012, 'Theatre as an Art of Emergence and Individuation', *Architectural Theory Review*, 17:1, 117–33
Oliver, Kelly, 2015, *Earth and World: Philosophy after the Apollo Missions* (New York: Columbia University Press)
Oliver, Kelly, 2018, 'Earth Love It or Leave It?', in Matthias Fritsch, Philippe Lynes and David Wood (eds), *Eco-Deconstruction: Derrida and Environmental Philosophy* (Fordham: Fordham University Press), pp. 339–54

Orozco, Lourdes, 2018, 'Animals in Socially-Engaged Performance Practices; Becomings on the Edge of Extinction', *Studies in Theatre and Performance*, 38:2, 176–89

O'Sullivan, Simon, 2010, 'Guattari's Aesthetic Paradigm: From the Folding of the Infinite/Finite Relation to Schizoanalytic Metamodelisation', *Deleuze and Guattari Studies*, 4:2, 256–86

Oswald, Alice and Paul Keegan, 2020, *Gigantic Cinema: A Weather Anthology* (London: Jonathan Cape)

Parker-Starbuck, Jennifer, 2011, *Cyborg Theatre: Corporeal/Technological Intersections in Multimedia Performance* (Basingstoke: Palgrave)

Patterson, Eddie, 2019, 'Compost and Air-Conditioning: Beyond Biospherical Performance and Toward the Shimmer', *Performance Research*: 'On Politics', 24:8, 31–6

Pavis, Patrice, 2003 [1996], *Analysing Performance: Theatre, Dance, and Film*, trans. David Williams (Ann Arbor: University of Michigan Press)

Pavis, Patrice, 2016, 'Watching the Spectator: New Perspectives on Spectating', in Christel Stalpaert, Katharina Pewny, Jeroen Coppens & Pieter Vermeulen (eds), *Unfolding Spectatorship: Shifting Political, Ethical, and Intermedial Positions* (Ghent: Academia Press), pp. 23–37

Pearson, Mike and Michael Shanks, 2001, *Theatre/Archaeology* (London: Routledge)

Perniola, Mario, 1995 [1990], *Enigmas: The Egyptian Moment in Art and Society*, trans. Christopher Woodall (London: Verso)

Peters, Julie Stone, 2002, 'Artaud in the Sierra Madres: Theatrical Bodies, Primitive Signs, Ritual Landscapes', in Elinor Fuchs and Una Chaudhuri (eds), *Land/Scape/Theatre* (Ann Arbor: University of Michigan Press), pp. 228–51

Peterson, Michael, 2018, 'Responsibility and the Non(bio)degradable', in Matthias Fritsch, Philippe Lynes and David Wood (eds), *Eco-Deconstruction: Derrida and Environmental Philosophy* (Fordham: Fordham University Press), pp. 249–60

Phelan, Peggy, 1993, *Unmarked: The Politics of Performance* (London: Routledge)

Phillips, Dana, 2003, *The Truth of Ecology: Nature, Culture and Literature in America* (Oxford: Oxford University Press)

Pignarre, Philippe and Isabelle Stengers, 2011 [2007], *Capitalist Sorcery: Breaking the Spell*, trans. Andrew Goffey (Basingstoke: Palgrave Macmillan)

Plato, 1998, *The Republic*, trans. Benjamin Jowett, https://www.gutenberg.org/files/1497/1497-h/1497-h.htm (accessed 19 August 2022)

Plato, 1999, *Ion*, trans. Benjamin. Jowett, https://www.gutenberg.org/files/1497/1497-h/1497-h.htm (accessed 19 August 2022)

Plumwood, Val, 1993, *Feminism and the Mastery of Nature* (London: Routledge)

Pradier, Jean-Marie, 2000, 'Animals, Angels, and Performance', *Performance Research*: 'On Animals', 5:2, 11– 22

Preece, Bronwyn, 2018, 'Environments, Ecologies, and Climates of Crisis: Engaging Disability Arts and Cultures as Creative Wilderness', in Bree Hadley and Donna McDonald (eds), *The Routledge Handbook of Disability Arts, Culture, and Media* (London: Routledge), pp. 281–94

Prigogine, Ilya and Isabelle Stengers, 2017 [1984], *Order out of Chaos: Man's New Dialogue with Nature* (London: Verso)

Quick, Andrew and Richard Rushton (eds), 2019, 'Introduction', *Performance Research*: 'On Theatricality', 24:3, 1–4

Rae, Paul, 2019, 'Archipelagic Performance: Scenes from Maritime Southeast Asia', *Theatre Journal*, 71:4, 455–73

Rancière, Jacques, 1995 [1992], *On the Shores of Politics*, trans. Liz Heron (London: Verso)

Raunig, Gerald, 2007 [2005], *Art and Revolution: Transversal Activism in the Long Twentieth Century*, trans. Aileen Derieg (Cambridge, MA: MIT Press)

Read, Alan, 1994, *Theatre & Everyday Life: An Ethics of Performance* (London: Routledge)

Read, Alan, 2008, *Theatre, Intimacy & Engagement: The Last Human Venue* (Basingstoke: Palgrave Macmillan)

Read, Alan, 2014, *Theatre in the Expanded Field: Seven Approaches to Performance* (London: Bloomsbury)

Read, Alan, 2020, *The Dark Theatre: A Book about Loss* (London: Routledge)

Readings, Bill, 1997, *The University in Ruins* (Cambridge, MA: Harvard University Press)

Reason, Matthew and Anja Lindelof (eds), 2016, *Experiencing Liveness in Contemporary Performance* (London: Routledge)

Rees, Sian, 2018, *Theatricalizing Dissent: An Examination of the Methodology and Efficacy of Performance in Contemporary Political Protest* (PhD dissertation, Goldsmiths, University of London)

Ridout, Nicholas, 2006, *Stage Fright, Animals, and Other Theatrical Problems* (Cambridge: University of Cambridge Press)

Roach, Joseph, 1993, *The Players Passion: Studies in the Science of Acting* (Ann Arbor: University of Michigan Press)

Robertson, Lisa, 2012, *Nilling: Essays on Noise, Pornography, the Codex, Melancholy, Lucretius, Folds, Cities and Related Aporias* (Toronto: Book Thug)

Robertson, Lisa, 2020, *The Baudelaire Fractal* (Toronto: Coach House Books)

Roffe, Jon and Hannah Stark (eds), 2015, *Deleuze and the Non/Human* (Basingstoke: Palgrave Macmillan)

Roumagnac, Vincent, 2020, *Reacclimatising the Stage (Skenomorphoses)* (Unpublished dissertation, University of Arts, Helsinki)

Rylance, Mark, 2021, 'Arts Should Tell Love Stories About Nature to Tackle the Climate Crisis', *Guardian*, 1 June, https://www.theguardian.com/culture/2021/jun/01/mark-rylance-arts-should-tell-love-stories-about-nature-to-tackle-climate-crisis (accessed 1 June 2022)

Said, Edward, 1978, *Orientalism* (New York: Pantheon Books)

Saldanha, Arun and Hannah Stark (eds), 2016, *Deleuze Studies*: 'A New Earth: Deleuze and Guattari in the Anthropocene', 10:4

Sartre, Jean-Paul, 1989 [1943], *Being and Nothingness: An Essay on Phenomenological Ontology*, trans. Hazel Barnes (London and New York: Routledge)

Savarese, Nicola, 2001, '1931: Antonin Artaud Sees Balinese Theatre at the Paris Colonial Exposition', *The Drama Review*, 45:3, 51–77

Schechner, Richard, 1968, '6 Axioms for Environmental Theatre', *The Drama Review*, 12:3, 41–63

Schechner, Richard, 1985, *Between Theatre & Anthropology* (Philadelphia: University of Pennsylvania Press)

Schechner, Richard, 1995, *The Future of Ritual: Writings on Culture and Performance* (London: Routledge)

Scheer, Edward (ed.), 2000, *100 Years of Cruelty: Essays on Antonin Artaud* (Sydney: Power Publications and Art Space)
Scheer, Edward, (ed.), 2003, *Antonin Artaud: A Critical Reader* (London: Routledge)
Schmidt, Theron, 2018, 'How We Talk About the Work Is the Work: Performing Critical Writing', *Performance Research*: 'On Writing & Performance', 23:2, 37–43
Schneider, Rebecca, 2015. 'New Materialisms and Performance Studies', *The Drama Review*, 59:4, 7–17
Schneider, Rebecca and Paul Rae, 2018, 'Extending a Hand / Lending an Ear', *Performance Research*: 'On Climate', 23:3, 13–24
Scholtz, Janae, 2015, *The Invention of a People: Heidegger and Deleuze on Art and the Political* (Edinburgh: Edinburgh University Press)
Schweitzer, Marlis and Joanne Zerdy (eds), 2014, *Performing Objects and Theatrical Things* (Basingstone: Palgrave Macmillan)
Sedgwick, Eve Kosofsky, 2003, *Touching, Feeling: Affect, Performativity, Pedagogy* (Durham, NC: Duke University Press)
Sellin, Eric, 1968, *The Dramatic Concepts of Antonin Artaud* (Chicago: University of Chicago Press).
Serres, Michel, 1995 [1990], *The Natural Contract*, trans. Elizabeth MacArthur and William Paulson (Ann Arbor: University of Michigan Press)
Serres, Michel, 2007 [1980], *The Parasite*, trans. Lawrence Schehr (Minneapolis: University of Minnesota Press)
Sessions, George, 1987, 'The Deep Ecology Movement: A Review', *Environmental Review*, 11:2, 105–25
Seymour, Nicole, 2018, *Bad Environmentalism: Irony and Irreverence in the Ecological Age* (Minneapolis: University of Minnesota Press)
Sharpe, Christina, 2016, *In the Wake: On Blackness and Being* (Durham, NC: Duke University Press)
Shaw, Robert, 2015, 'Bringing Deleuze and Guattari Down to Earth through Gregory Bateson', *Theory, Culture & Society*, 32: 7–8, 151–71
Sheldon, Rebekah, 2015, 'Form/Matter/Chora: Object-Orientated Ontology and Feminist New Materialism', in Richard Grusin (ed.), *The Non-Human Turn* (Minneapolis: University of Minnesota Press), pp. 193–222
Sim, Stuart, 2001, *Lyotard and the Inhuman* (Cambridge: Icon Books)
Simondon, Gilbert, 2009, 'The Position of the Problem of Ontogenesis', trans. Gregory Flanders, *Parrhesia*, 7, 4–16
Simondon, Gilbert, 2017 [1958], *On the Mode of Existence of Technical Objects* (Minneapolis: Univocal Press)
Simondon, Gilbert, 2020 [2005], *Individuation: In Light of Notions of Form and Information*, trans. Taylor Adkins (Minneapolis: University of Minnesota Press)
Sloterdijk, Peter, 2009, *Terror from the Air*, trans. Amy Patton and Steve Corcoran (Los Angeles: Semiotext(e))
Sloterdijk, Peter, 2011, *Bubbles: Spheres 1*, trans. Wieland Hoban (Los Angeles: Semiotext(e))
Slovic, Scott and Paul Slovic (eds), 2015, *Nerves and Numbers: Information, Emotion and Meaning in a World of Data* (Corvallis: Oregon State University Press)
Smithson, Robert, 1996, *Robert Smithson: The Collected Writings*, ed. Jack Flam (Berkeley and Los Angeles: University of California Press)

Sofer, Andrew, 2003, *The Stage Life of Props* (Ann Arbor: University of Michigan Press)

Sofer, Andrew, 2013, *Dark Matter: Invisibility in Drama, Theater and Performance* (Ann Arbor: University of Michigan Press)

Sontag, Susan, 1976, 'Artaud: An Essay by Susan Sontag', in Susan Sontag (ed.), *Antonin Artaud: Selected Writing* (Berkeley: University of California Press), pp. xvii–lix

Sophocles, 1986, *Plays: 1*, trans. Don Taylor (London: Methuen)

Spivak, Gayatri Chakravorty, 2014, 'Planetarity', in Barbara Cassin (ed.), *Dictionary of Untranslatables: A Philosophical Lexicon* (Princeton: Princeton University Press), p. 1223

Spreen, Constance, 2003, 'Resisting Plague: The French Reactionary Right and Artaud's Theater of Cruelty', *Modern Language Quarterly*, 64:1, 71–96

Stalpaert, Christel, 2018, 'Cultivating Survival with Maria Lucia Cruz Correia: Towards an Ecology of Agential Realism', *Performance Research*: 'On Climate', 23:3, 44–55

Stalpaert, Christel, 2020, 'Figures Performing Prototypes of Composite Bodies: Kris Verdonck's ontological Politics of Time and Movement', in Peter Eckersall and Kristof van Baarle (eds), *Machine Made Silence: The Art of Kris Verdonck* (Aberystwyth: Performance Research Books), pp. 118–31

Stalpaert, Christel, Katharina Pewny, Jeroen Coppens and Pieter Vermeulen (eds), 2016, *Unfolding Spectatorship: Shifting Political, Ethical, and Intermedial Positions* (Ghent: Academia Press)

Stalpaert, Christel, Kristof Van Baarle and Laura Karreman (eds), 2021, *Performance and Posthumanism: Staging Prototypes of Composite Bodies* (Cham: Palgrave Macmillan)

Standing, Sarah Ann, 2019, 'Re-Visit/Re-Examine/Re-Contextualise/Re-Ignite: Protest and Activism as Performance', in Peter Eckersall and Helena Grehan (eds), *The Routledge Companion to Theatre and Politics* (London: Routledge), pp. 341–4

Stengers, Isabelle, 2010, *Cosmopolitics*, trans. Robert Bononno (Minneapolis: University of Minnesota Press)

Stengers, Isabelle, 2012, 'Reclaiming Animism', *E-Flux Journal* 36#, https://www.e-flux.com/journal/36/61245/reclaiming-animism (accessed 22 June 2022)

Stengers, Isabelle, 2015 [2009], *In Catastrophic Times: Resisting the Coming Barbarianism*, trans. Andrew Goffey (Lüneberg: Open Humanities Press)

Stephens, Elizabeth and Annie Sprinkle, 2012, 'On Becoming Appalachian Moonshine', *Performance Research*: 'On Ecology', 17:4, 61–6

Stern, Daniel, 2018 [1985], *The Interpersonal World of the Infant: A View from Psychoanalysis and Developmental Psychology* (London: Routledge)

Stevens, Lara, 2020, 'Assembling Non-Presence in *The Aborigine Is Present*', in Peter Eckersall and Helena Grehan (eds), *The Routledge Companion to Theatre and Politics* (London: Routledge), pp. 98–101

Stevens, Lara, Peta Tait and Denise Varney (eds), 2018, *Feminist Ecologies: Changing Environments in the Anthropocene* (Cham: Palgrave Macmillan)

Stewart, Kathleen, 2007, *Ordinary Affects* (Durham, NC: Duke University Press)

Stoekl, Allan, 2007, *Bataille's Peak: Energy, Religion, and Postsustainability* (Minneapolis: University of Minnesota Press)

Su, Tsu-Chung, 2012, 'Artaud's Journey to Mexico and His Portrayals of the Land', *CLCWEB, Comparative Literature and Culture*, 14:15, 1–8, https://doi.org/10.7771/1481-4374.2151 (accessed 21 June 2022)
Sutil, Nicolás Salazar, 2018, *Matter Transmission: Mediation in a Paleocyber age* (London: Bloomsbury)
Szerszynski, Bronislaw, 2018, 'Drift as a Planetary Phenomenon', *Performance Research*: 'On Drifting', 23:7, 136–44
Tait, Peta, 2012, *Wild and Dangerous Performances: Animals, Emotions, Circus* (Basingstoke: Palgrave Macmillan)
Tait, Peta, 2018, 'Performing Ghosts, Emotions, and Sensory Environments', in Lara Stevens, Peta Tait and Denise Varney (eds), *Feminist Ecologies: Changing Environments in the Anthropocene* (Cham: Palgrave Macmillan), pp. 175–91
Taylor, Diana, 2003, *The Archive and the Repertoire: Performing Cultural Memory* (Durham, NC: Duke University Press)
Thomas, Keith, 1984, *Man and the Natural World: Changing Attitudes in England 1500–1800* (London: Penguin)
Thompson, James, 2009, *Performance Affects: Applied Theatre and the End of Effect* (Basingstoke: Palgrave Macmillan)
Tian, Min, 2018, *The Use of Asian Theatre for Modern Western Theatre: The Displaced Mirror* (Cham: Palgrave Macmillan)
Tiqqun, 2010 [2001], *The Cybernetic Hypothesis*, trans. Robert Hurley (Cambridge, MA: MIT Press)
Toadvine, Ted, 2018, 'Thinking after the World: Deconstruction and Last Things', in Matthias Fritsch, Philippe Lynes and David Wood (eds), *Eco-Deconstruction: Derrida and Environmental Philosophy* (Fordham: Fordham University Press), pp. 50–80
Todd, Zoe, 2016, 'An Indigenous Feminist's Take on the Ontological Turn: "Ontology" is Just Another Word for Colonialism', *Journal of Historical Sociology*, 29, 4–22
Tompkins, Joanne, 2020, 'Re-Readings: Staging Place: The Geography of Modern Drama by Una Chaudhuri', *Contemporary Theatre Review*, 2020:3, 408–9
Tsing, Anna Lowenhaupt, 2015, *The Mushroom at the End of the World: On the Possibility of Life in Capitalist Ruins* (Princeton: Princeton University Press)
Tsing, Anna, Nils Bubandt, Elaine Gan and Heather Anne Swanson (eds), 2017, *Arts of Living on a Damaged Planet: Ghosts and Monsters of the Anthropocene* (Minneapolis: University of Minnesota Press)
Tuana, Nancy, 2008, 'Viscous Porosity: Witnessing Katrina', in Stacy Alaimo and Susan Hekman (eds), *Material Feminisms* (Bloomington: Indiana University Press), pp. 188–213
Twitchin, Mischa, 2016, *The Theatre of Death: The Uncanny in Mimesis: Tadeuz Kantor, Aby Warburg, and an Iconology of the Actor* (Basingstoke: Palgrave Macmillan)
Twitchin, Mischa and Carl Lavery, 2019, 'Introduction', *Performance Research*: 'On Animism', 24:6, 1–5
Van Baarle, Kristof, 2018, *From the Cyborg to the Apparatus: Figures of the Posthuman in The Philosophy of Giorgio Agamben and Performing Arts of Kris Verdonck* (PhD dissertation, University of Ghent)

Van Baarle, Kristof, 2020, 'End-Time Attitudes', in Peter Eckersall and Kristof van Baarle (eds), *Machine Made Silence: The Art of Kris Verdonck* (Aberystwyth: Performance Research Books), pp. 152–64

Van der Tuin, Iris, 2014, 'Diffraction as a Methodology for Feminist Onto-Epistemology: On Encountering Chantal Chawaf and Posthuman Interpellation', *Parallax*, 20:3, 231–44

Van Dooren, Thom, 2014, *Flight Ways: Life and Loss at the Edge of Extinction* (New York: Columbia University Press)

Van Kerkhoven, Marianne and Anoek Nuyens (eds), 2012, *Listen to the Bloody Machine: Creating Kris Verdonck's End* (Utrecht: Utrecht School of the Arts)

Varela, Francisco, 1999, *Ethical Know-How: Action, Wisdom, Cognition* (Stanford: Stanford University Press)

Varney, Denise, 2018, 'Climate Change Guardian Angels: Feminist Ecology and the Activist Tradition', in Lara Stevens, Peta Tait and Denise Varney (eds), *Feminist Ecologies: Changing Environments in the Anthropocene* (Basingstoke: Palgrave Macmillan), pp. 135–53

Vasudevan, Pavithra, 2012, 'Performance and Proximity: Revisiting Environmental Justice in Warren County, North Carolina', *Performance Research*: 'On Ecology', 17:4, 18–26

Vinge, Vernor, 1993, 'The Coming Technological Singularity: How to Survive in the Posthuman Era', https://frc.ri.cmu.edu/~hpm/book98/com.ch1/vinge.singularity.html (accessed 21 June 2022)

Visser, Joeri, 2021, *Antonin Artaud and the Healing Practices of Language: How Life Matters in Artaud's Later Writings* (London: Bloomsbury)

Viveiros de Castro, Eduardo, 2016, *The Relative Native: Essays on Indigenous Conceptual Worlds* (Chicago: Hau Books)

Von Mossner, Alexa Weik, 2017, *Affective Ecologies: Emotion, Empathy, and Environmental Narrative* (Columbus: Ohio State University Press)

Wade, Sarah, 2020, 'From the Sublime to the Ridiculous: Extinction in the Work of Marcus Coates', *Performance Research*: 'On Disappearance', 24:7, 37–42

Weber, Samuel, 2004, *Theatricality as Medium* (Fordham: Fordham University Press)

Weiss, Allen S., 1992, 'Radio, Death, and the Devil: Artaud's *Pour en finir avec le Jugement de Dieu*', in Douglas Kahn and Gregory Whitehead (eds), *Wireless Imagination: Sound, Radio, and the Avant-Garde* (Cambridge, MA: MIT Press), pp. 269–307

Welton, Martin, 2012, *Feeling Theatre* (Basingstoke: Palgrave Macmillan)

Welton, Martin, 2018, 'Making Sense of Air: Choreography and Climate in *Calling Tree*', *Performance Research*: 'On Climates', 23:3, 80–90

Wickham, Glynne, 1994, *A History of the Theatre*, 2nd edn (London: Phaidon)

Wickstrom, Maurya, 2019, 'Wet Ontology, Moby-Dick, and the Oceanic in Performance', *Theatre Journal*, 71:4, 475–91

Williams, David, 2000, 'The Right Horse, the Animal Eye – Bartabas and Théâtre Zingaro', *Performance Research*: 'On Animals', 5:2: 29–49

Williams, David, 2006, 'Weather', *Performance Research*: 'A Lexicon', 11:3, 142–4

Williams, Raymond, 1988, *Keywords: A Vocabulary of Culture and Society*, revised and expanded edn (London: Fontana)

Wilmer, Stephen E. and Karen Vedel (eds), 2020, *Nordic Theatre Studies*: 'Theatre and the Anthropocene', 32:1

Wilson, Elizabeth A., 2015, *Gut Feminism* (Durham, NC: Duke University Press)

Wilson, Louise Ann, 2019, 'Dorothy Wordsworth and Her Female Contemporaries: A Feminine 'material' Sublime Approach to the Creation of Walking Performance in Mountainous Landscapes', *Performance Research*: 'On Mountains', 24:2, 109–19

Wolfe, Cary, 2010, *What Is Posthumanism?* (Minneapolis: University of Minnesota Press)

Wood, David, 2018, 'The Eleventh Plague: Thinking Ecologically after Derrida', in Matthias Fritsch, Philippe Lynes and David Wood (eds), *Eco-Deconstruction: Derrida and Environmental Philosophy* (Fordham: Fordham University Press), pp. 29–49

Wood, Sarah, 2014, *Without Mastery: Reading and Other Forces* (Edinburgh: Edinburgh University Press)

Woodward, Ashley, 2016, *Lyotard and the Inhuman Condition: Reflections on Nihilism, Information and Art* (Edinburgh: Edinburgh University Press)

Worthen, Hana, 2020, *Humanism, Drama, and Performance: Unwriting Theatre* (Basingstoke: Palgrave Macmillan)

Woynarski, Lisa, 2015, 'Locating an Indigenous Ethos in Ecological Performance', *Performing Ethos*, 5, 17–30

Woynarski, Lisa, 2020, *Ecodramaturgies: Theatre, Performance and Climate Change* (Basingstoke: Palgrave Macmillan)

Wynter, Sylvia, 1971, 'Novel and History: Plot and Plantation', *Savacou*, 1:5, 95–102

Yarrow, Ralph, 2001, 'Theatre Degree Zero', *Studies in the Literary Imagination*, 34:2, 75–92

Young, Harvey, 2010, *Embodying Black Experience: Stillness, Critical Memory and the Black Body* (Ann Arbor: University of Michigan Press)

Yusoff, Kathryn, 2017, *A Billion Black Anthropocenes or None* (Minneapolis: University of Minnesota Press)

Zimmerman, Lee, 2020, *Trauma and the Discourse of Climate Change: Literature, Psychoanalysis and Denial* (London: Routledge)

Index

abhumanism 20, 128, 141–7, 157, 166, 172, 174n.3–7, 176n.15, 213
activism ix–xvi, 45–6, 56–7, 115, 155, 211
actual 7, 13, 94, 97–101, 107–10, 118n.7, 131, 136, 163, 193, 213, 218
Agamben, Giorgio xii, 47, 48, 63, 74n.45, 176n.21
agential realism 67, 107, 110, 118n.1, 119n.10
Alaimo, Stacey xiii, 84, 87, 89
alchemy 154, 155, 161, 163, 164–73, 177n.26
anamnesis 13, 185, 198, 204n.8
animism 115, 155, 161, 163, 164–73, 180n.46–9, 205n.17, 213
Anthropocene vi, xvn.12, 6, 26–32, 49, 53, 66, 75n.52–3, 84, 92, 103n.16, 111, 149, 152, 202, 213, 214, 216
anthropos 13, 50, 54, 56, 68n.2, 81, 147, 149, 208, 214
archipelagic spectating 190–203
archipelago 20, 21, 181–203, 211, 214
Arons, Wendy 29
Artaud, Antonin xii, 1, 2, 11, 20, 21, 21n.1, n.5, 25–6, 67, 95, 96, 105n.32, 109, 118n.6, 135, 136, 141–81, 183, 186, 187, 189, 190, 193, 194, 199, 200, 205n.17, n.18, 208, 209–10, 214

Barad, Karen xiii, 4, 10, 19, 20, 67, 71n.20, 75n.54, 106–10, 111, 112, 113, 116, 117, 117n.2, 118n.4–6, 119n.8, n.10–11, n.13, 120n.15, 122, 131, 132, 134, 163, 169–70, 180n.49, 183, 196, 210, 214, 215
Bataille, Georges 10, 90, 99, 105n.35, 174n.2
Bateson, Gregory x, 8, 23n.14, 40, 42, 56, 61, 72n.29, n.31, 73n.40, 80
Battista, Silvia 60, 65, 117n.2, 120n.15
Beckett, Samuel xii, 33, 96, 109, 119n.9, 142, 147, 197, 211
Bennett, Jane 56, 75n.54, 84
Berardi, Franco x, 205n.13
Bergthaller, Hans 25, 35, 51
Braidotti, Rosa 19, 84, 86, 89, 102n.12
Brassier, Ray 48, 87–9, 104n.25
Brecht, Bertolt 20, 52, 108, 112, 119n.11, 129, 133, 137n.1, 183, 204n.5, 215

Carson, Rachel 156
Carter, Paul 191–2, 210
capitalist sorcery 6, 149–58, 163, 171, 173, 178n.34, 210
Capitalocene xi, xvn.11, 6, 22n.9, 214
Caygill, Howard 160, 62
Chakrabarty, Dipesh, 81, 102n.7
Chaudhuri, Una 18, 30, 31, 32, 32–8, 39, 40, 42, 43, 50–4, 58, 59, 61, 65, 67, 69n.11, n.13, 70n.16, 71n.20–1, 75n.52, 173n.2
Chen, Mel 81, 89
Clark, Nigel 14, 180n.47
Clifford, James 151, 152, 187n.27

Colebrook, Claire 19, 77n.66, 81–2, 83, 90, 95, 103n.20, 127, 143, 144, 145, 161, 205n.19, 208, 217
correlationism 88–9, 214
Corvin, Michel 10, 20, 128, 141–7, 157, 166, 172, 174n.3–7, 175n.8, n.10, 213
cosmic earth 4, 11, 13–15, 124, 146, 163, 167, 192, 205n.16, 217
Craig, Edward Gordon 142, 144, 145, 166, 175n.11, 176n.15
cruel ecology 182–206
cruelty 2, 11, 19, 21, 112, 136, 146, 147, 149–57, 158, 160, 162, 166, 170, 171, 173–4n.2, 175n.13, 176n.16, n.18, 181n.52, 183, 185, 193, 195, 199, 208, 214
Cull, Laura 62, 163
cybernetics 8, 9, 23n.13, 42, 50, 56, 68n.1, 72n.28, 73n.40, 76n.59, 80–1, 130, 137n.3, 139n.13
cyborg 62, 84–5, 102n.14, 109

Danowski, Déborah and Eduardo Viveiros de Castro 56, 88–9, 103n.16, 104n.23, n.26, n.28, 178n.31
decolonial ecology 57, 79
DeLanda, Manuel 84, 192
Deleuze, Gilles 7, 8, 22n.6, n.10, 140n.19
Deleuze, Gilles and Félix Guattari ix, 3, 13–16, 22n.7, 23n.21, 24n.22, n.27, 65, 73n.37, 115, 133, 159, 167, 179n.40, 180n.43, n.46, 192, 200, 208, 209, 214
Demaitre, Ann 164
de-phasing 13, 19, 94, 97, 98, 100, 101, 115, 124, 132, 135, 136, 169, 188, 197, 209, 217
Derrida, Jacques xii–iii, xvn.14, xvin.15, 20, 49, 68n.6, 73n.42, 120n.15, 157, 158–64, 166, 167, 169, 173, 178n.35–6, 179n.37–8, n.41, 180n.42
deterritorialisation 14–15, 24n.22, 114, 127, 133, 214
diffraction 106–10, 118n.4–5, 131, 166, 198, 214

earth 1–9, 12–16, 27, 37, 41–2, 58, 59, 63, 67, 82–3, 86, 88, 91, 94, 99–101, 111–12, 114n.19, 122–40, 145–6, 149, 153, 161–2, 168, 171–2, 188, 190, 192, 202, 203, 208, 214, 216
ecocriticism xivn.2, 1, 18, 28, 32, 68n.7, 78–105, 187, 207–11
ecodramaturgy 29–30, 53, 61
ecologising theatre 17, 19–20, 94, 122–40, 182, 214–15
ecomimesis 17, 24n.26, 61
enigma 96, 134, 178, 195, 197
Environmental Humanities 2, 17, 19, 78–105, 128, 207–8, 215
Epstein, Jean 94
Etchells, Tim 76n.2, 204n.11

Fensham, Rachel 4, 182, 203n.2, 204n.11
Féral, Josette 20, 130–2, 134, 139n.14–15, 140n.16
Ferdinand, Malcom 57, 76n.57, 79, 106, 168, 210
Fischer-Lichte, Erica 20, 129–30, 132, 139n.13
Fuchs, Elinor 31, 43, 71n.24, 72n.32
Fukuyama, Francis 85, 103n.19

gestus 20, 21, 96, 183–9, 194, 197, 198, 199, 204n.5–6, 215
Glissant, Édouard 2, 4, 5, 16, 17, 21, 24n.25, 101, 106, 111, 183, 189, 190–4, 195, 199, 200, 204n.3, 205n.15, 210, 211, 217
Goodall, Jane 123, 149–50, 176n.22, 178n.35
Gotman, Kélina 189
Grosz, Elizabeth xi, 3, 13, 98, 196
Guattari, Félix ix–xvi, 3, 5, 6–7, 14–16, 23n.21, 24n.22, n.27, 65, 66, 73n.37–40, 106, 113–17, 122, 131, 133, 136, 138n.10, 147, 159, 163, 167, 170, 171, 172, 179n.40, 180n.43, 188, 192, 197, 200, 208, 209, 210, 214, 218

Haraway, Donna 6, 19, 22n.9, 84, 89, 90, 91, 92, 93, 102n.14, 104n.21, n.27, 140n.18, 150, 177n.25, 199
Harney, Stefano and Fred Moten 16, 66
Heidegger, Martin 12, 14, 15, 24n.22–4, 123, 137n.2
Hudson, Julie 29–30, 54, 68n.7, 69n.10, 71n.21, 76n.59, 102n.6
humanism 16, 61, 62, 68n.2, 77n.66, 80–91, 122–8, 137n.1–2, 141–7, 157, 173, 208, 215

im/materialism xiii, 9, 13, 17, 66, 67, 94, 95, 96, 98, 126, 131, 132, 136, 140n.16, 148, 162, 166, 192, 194, 196, 198, 209, 215
Indigeneity 29, 30, 56, 79, 80, 84, 85, 89, 92, 104n.28, 111, 149–57, 178n.31, 180n.48
individuation 4, 17, 19, 20, 97–101, 105n.33–4, 131, 136, 147, 165, 175n.8–9, 179n.40, 188, 209, 215, 217
inhuman 12, 96, 119n.14, 145, 217
intra-activity 17, 80, 106–10, 169, 207, 215–16
Iovino, Serenella and Serpil Oppermann 91–2

Jannarone, Kimberly 150–1, 152, 159, 176n.24, 177n.26, 178n.30, n.34–5

Kelleher, Joe xivn.5, 73n.41, 189, 197
Kermode, Frank 93–4
Kershaw, Baz 8–9, 18, 22n.13, 30, 31, 32, 40, 42–6, 48, 49, 50–1, 54–9, 61, 64, 65, 67, 69n.10–11, n.13, 72n.31–4, 73n.36–40, 74n.46, 76n.59–60, 80, 139n.13, 173n.2, 186
Kirby, Vicky 19, 85, 89

Lacan, Jacques 114, 121n.19–20, 138n.9, 181n.50
Latour, Bruno xviii.15, 25, 26, 28, 34, 74n.51, 82, 102n.9, 168, 176n.20, 180n.48

Lavery, Carl 52, 70n.17, 71n.25, 105n.31, 180n.47, 206n.23
Lehmann, Hans-Thies 74n.43, 95, 96
LeMenager, Stephanie 92
Lynes, Philippe 12
Lyotard, Jean-François xiii, 12, 19, 20, 21, 90, 96, 105n.30, 110–13, 116, 117, 111n.12–14, 126, 132, 135, 138n.7, n.9, 145, 146, 161, 163, 170, 171, 182–90, 193–4, 195, 197, 203, 204n.5, 205n.6–8, n.11, 208, 210, 215, 216

Manning, Erin 3, 4, 28, 64, 105n.33, 182, 206n.24, 208
Marder, Michael 12, 23n.17, 69n.9, 159–60
Margulis, Lynn 150, 177n.25
Marranca, Bonnie 30, 32, 38–42, 43, 48, 50, 51, 54, 60–5, 67, 71n.24–6, 186
May, Theresa J. 29, 52, 53, 61
Mbembe, Achille 16, 155, 200, 210
McKenzie, Jon 23n.13, 42, 118n.7, 137n.3
McKittrick, Katherine 22n.9, 217
Meillassoux, Quentin 87–8, 89, 104n.25
methods 25–77
Morton, Timothy 6, 17, 23n.13, 24n.26, 40, 41, 61, 62

Nail, Thomas 83
new materialism 6, 13, 20, 29, 55–6, 75n.54, 80, 84
Neyrat, Frédéric 155, 162
non-human xiii, 2, 3, 11, 12, 13, 15, 17, 20, 24n.27, 58, 60–4, 76n.60, 87–90, 101, 104n.27, 110–13, 122–8, 132–6, 142–7, 215

oikos xiii, 12, 36, 38, 79, 90, 110–13, 117, 145, 146, 160, 171, 184, 187, 190, 196, 198, 202, 211, 216
opacity 5, 17, 47, 95, 111, 144, 147, 170, 195, 211
orientalism 150–2, 178n.28

passibililty 112, 197, 216
Pavis, Patrice 22n.7, 183, 204n.4, n.11
performance analysis 182–90
Perniola, Mario 195
pharmacological 27, 28, 68n.6, 216
pharmakon 27, 28, 50, 185, 208, 216
Pignarre, Philippe xivn.1, 6, 154–5, 171
planetary 6, 7, 12–17, 31, 56, 81, 88, 101, 117, 136, 146, 151–8, 173, 188–94, 207–11
plantationocene xi, xvn.11, 6, 22n.9, 199, 216–17
Plato 7, 19–20, 124–28, 129, 130, 131, 138n.8, 146, 164, 166, 174n.6
Plumwood, Val 41–2, 71n.27
postcolonialism 27, 56–7, 76n.57, 88, 149–57, 192, 210, 216
posthumanism 84–91, 122–30, 140–7, 217
pre-individual 3, 5, 19, 76n.60, 90, 97–101, 105n.34, 127, 128, 135, 144, 147, 162, 175n.13, 192, 196, 197, 200, 209, 210, 216, 217

quantum ecology 107–10, 188n.5, 169
queer ecology 53, 61, 79, 81, 102n.5, 107, 137n.1

Rae, Paul xiii, 192
Read, Alan 18, 30, 31, 47–50, 60–7, 108, 123, 137n.3, 138n.4
refrains 196, 199
relationality 9, 10, 21n.2, 66, 82, 107, 166, 208
Robertson, Lisa 58, 91, 186, 187–8, 190, 195, 204n.10, 205n.13

Schechner, Richard 3, 73n.35, 123, 154
Schneider, Rebecca xiii, 56
Serres, Michel 83, 86, 102n.10
Sharpe, Christina 199, 210
Sheldon, Rebekah 89–90, 95, 118n.3
Simondon, Gilbert 3, 4, 19, 21n.3, 78, 97–101, 105n.33–4, 106, 117n.1, 124, 131, 162, 166, 168, 192, 208, 209, 210, 215, 217
site-specifics 30, 43, 45, 54–60, 74n.48

Sloterdijk, Peter 54, 151–2
Smithson, Robert 125–6, 138n.5–6
speculative realism 67, 87–8, 118n.3
Stalpaert, Christel xvn.14, 60, 102n.11, 175n.14
Stengers, Isabelle xii, 6, 27, 83, 152, 154–5, 168, 171, 180n.46, n.48, 205n.17
storytelling 91–5, 140n.18, 143
sympoiesis 84
Szerszynski, Bronislaw 12, 191

terrestrial 217
theatre ecology 1–7, 8–10
theatricalising ecology 95–101, 218
theatricality 95–101, 129–36, 218
transference 19, 66, 114–16, 120n.18, 171, 187, 196, 205n.12
transversality ix, 45, 66, 73n.38, 113–17, 120n.17–18, 127, 131, 172, 188, 197, 205n.13, 208, 210, 218
Tsing, Anna Lowenhaupt xivn.1, 6, 84, 85, 92–3, 199

Van Doreen, Thom 48, 93–4
Van Kerkhoven, Marianne 202, 205n.22
veer ecology 66, 77n.67
Verdonck, Kris 128, 175n.14, 200, 201–2, 205n.22, 206n.23, 210, 211
Vienne, Gisèle 96, 128, 142, 200, 203, 210
virtuality xii, 7–8, 67, 71n.20, 73n.39, 76n.60, 95–101, 106–10, 114–17, 128–36, 140n.16, 148, 162–73, 182, 187, 189, 190, 193, 209, 218
Visser, Joeri 194, 205n.18
Von Mossner, Alexa Weik 92–4

Weber, Samuel 13, 20, 132–6, 142, 210
Wood, Sarah 189, 198
Worthen, Hana 10, 123, 137n.1, 140n.17, 175n.9
Woynarski, Lisa xivn.3, 29–30, 68n.7, 76n.55
writing 64–7, 189–94

Yusoff, Kathryn 27, 85, 168, 217

EU authorised representative for GPSR:
Easy Access System Europe, Mustamäe tee 50,
10621 Tallinn, Estonia
gpsr.requests@easproject.com

www.ingramcontent.com/pod-product-compliance
Ingram Content Group UK Ltd.
Pitfield, Milton Keynes, MK11 3LW, UK
UKHW021833210426
5322IPUK00004B/166